# DIGITAL LOGIC
# RTL & VERILOG
## Interview Questions

# A Practical Study Guide
# for Design Engineers

## VERILOGCODE.COM

Ordering Information:
Quantity sales. Special discounts are available on quantity purchases by corporations,
associations, and others. Orders by U.S. trade bookstores and wholesalers, please visit:
www.VerilogCode.com


Printed in the United States of America

First Printing, May 2015

Revision 1.6 August 2016

ISBN-13: 978-1512021462

ISBN-10: 1512021466

www.VerilogCode.com


For permission requests, contact VerilogCode.com

# DIGITAL LOGIC
## RTL & VERILOG
## Interview Questions

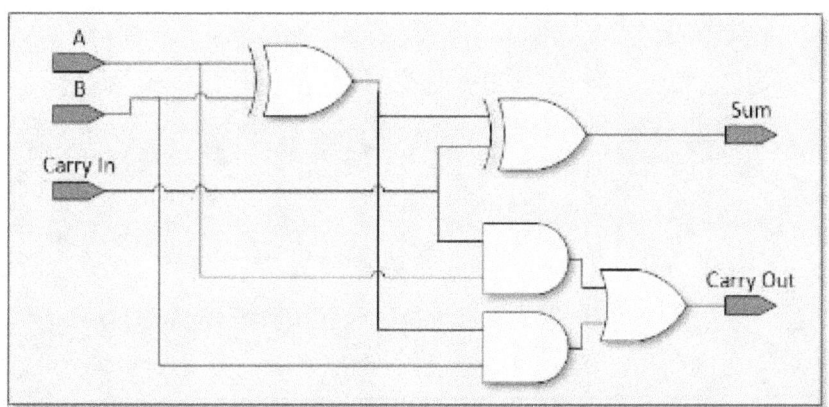

# A Practical Study Guide
# for Design Engineers

## VERILOGCODE.com

# DIGITaL LOGIC
# RTL & VeriLOG
# Interview Questions

**About the Author:**

Trey Johnson has been designing digital logic circuits and writing RTL code in both Verilog and VHDL languages for almost twenty years.

In the late 1990's, Johnson designed and developed some of the first multimedia hardware components used inside early smartphones, with his primary design focus in video and graphical subsystems, LCD and camera subsystems, and 2D hardware accelerators. He has worked closely and designed hardware components for both ARM and DSP processors and cache subsystems. He also has experience designing I/O peripherals such as resistive touch screen displays, magnetic card readers, PCI Express controllers and SerDes subsystems, and memory controllers.

Johnson has been granted three United States Patents for his digital design solutions. He is the founder of **VerilogCode.com** which is a website dedicated to sharing information about Verilog and RTL design.

Please visit the website for more digital design and job interview questions and to also share your own experiences.

This book is dedicated to Brandi, Tucker, Gunner and Alexa

*Thank you for riding with me on life's waves of change,*

*...and for embracing the journey along the way.*

# Table of Contents

# List of Questions

### RTL Verilog Syntax Questions

1. Explain *blocking* versus *non-blocking* statements
2. Show Verilog code for *bitwise* versus *conditional* operators
3. Verilog code for logic gates: *and, or, nand, nor, xor, xnor*
4. Verilog code for bitwise reduction
5. Verilog code multiplying and dividing by powers of 2
6. Verilog code for sign extension and concatenation
7. Write Verilog Code for *asynchronous* and *synchronous* flip flops
8. Verilog coding - what are three ways to code a mux
9. What type of circuit would the synthesis tool create for mux code
10. Verilog code for latch versus flip flop and draw timing diagram

### RTL Logic Design Questions

11. Design a circuit to detect if a signal transitions in any direction
12. Design a circuit to detect a 1 cycle high pulse (synchronously)
13. Design a sequence detector circuit to detect 1,0,1,1,0 using FSM
14. Verilog code to detect a pattern 10110 anywhere in last 8 samples
15. For the timing diagram show, write Verilog code to create it
16. Design a debounce circuit to remove input glitches
17. Write Verilog code to convert BCD to gray code
18. Design a synchronous fifo module using dual port RAM
19. Design a circuit to detect if number is divisible by three
20. Design a circuit to calculate Fibonacci sequence
21. Design a circuit to find the maximum and second highest number
22. Design a circuit to output second, minute, and hour from 1ms input
23. Given the timing diagram, write the equivalent Verilog code
24. Draw the structure of a digital FIR filter with 5-taps

### Clock Dividers, Clock Gating, and Reset Questions

25. Design a clock divide by 2 circuit
26. Design a clock divide by 3 circuit (with 50 percent duty cycle)
27. Design a clock divider by *N* circuit (with 50 percent duty cycle)
28. Design a glitch free clock gating cell with enable pin
29. How to detect a rising edge of an input signal if clocks are off ?
30. Design a reset circuit with async assertion and sync deassertion

## Clock Domain Crossing Questions

31. What is metastability?
32. Design a circuit from a slow clock domain to a fast clock domain
33. Design a circuit to handle CDC from fast domain to a slow domain
34. How would the circuit change if you need to synchronize a bus?
35. Gray coding techniques to cross clock domains

## Power Related Questions

36. Describe two components of power
37. Describe how to reduce static power
38. Describe how to reduce dynamic power
39. Describe low power RTL coding techniques

## Refresher: Digital Logic Questions

40. What is definition of setup and hold time for a flip flop?
41. Venn Diagram and Boolean Logic
42. Logic Gate Design - Transistor Level
43. Cross section of a transistor and process nodes
44. Karnaugh Maps (clk divide by 3 circuit not 50-50 duty cycle)
45. Half Adder using *xor* gates for addition and an *and* gate for carry
46. Using only 2 input muxes, create a *nand, nor, inverter, or, and, xor*
47. How to use and an *xor* gate like a controlled inverter?
48. Design an *inverter, and , or* and *xor* gate using just *nand* gates
49. Create 4:1 mux using 2:1 mux
50. What is the fastest frequency this circuit can run?
51. Convert this decimal value to binary, hex, and octal format

## Logical Thinking Questions

52. Four Gallons of Water
53. The Path to Freedom
54. Three Light Switches
55. Multiplication Question
56. Einstein's Riddle

# List of Figures

# List of Verilog Coding Examples

# List of Tables

# Introduction

I'm 30,000 feet above the ground, on a plane headed to San Diego for a job interview. I'm a little anxious, and I take some comfort in looking out of the small window pane next to me. My eyes wander back down to the magazine article sitting on my lap, then I read these words:

*"Transitions. That's all life is, and it's tougher than physics. From school to work to retirement to dead"*

The magazine article is about children who are interviewing for preschool, and the most important characteristic that the administrators look for is how well can the child *adapt* to change with new surroundings and new rules. But this article could have been written about me: an engineer who spent eighteen years working for the same company, and one day was suddenly let go as a result of a corporate downsizing. I am now the one who must adapt to change, new surroundings, and new rules.

This book documents real interview questions that I encountered from my own personal job interviewing experiences with some of the top-tier semiconductor companies in the world. This book also contains fundamental digital design material and practical Verilog code examples that I created based on the themes from the types of questions that I experienced first hand. This book will help prepare you for your own interviewing process. It is by no means the end-all, but rather, consider this book as a great starting point. As you read through some of the questions, I will also share with you some of my personal insight and knowledge in the *Author's Tips* section, which I have acquired through my career as a digital logic professional.

Interviewing for a job is like going on a date; at first you may feel a little nervous or awkward, but after some time and more interviews you soon become more comfortable and confident. Do not get discouraged in the beginning! The job interviewing experience can be daunting. It will test your mental toughness. I experienced headaches during my first few interviews because of the long hours of mental stress. But the saying is true that *practice makes perfect*. After several more interview attempts, I became more comfortable, developed a sense of calmness, and felt more prepared to answer the questions.

I've encountered many different types of interview questions ranging from real world practical examples, to academic textbook or theoretical questions (usually asked by people with PHDs with not much practical experience), to tricky questions using some obscure circuit (which would never be applicable in the real world), to behavioral questions (usually asked by Human Resource representatives). This book focuses on real world practical examples, and it also discusses some of the tricky and obscure questions that are asked. Preparing for behavioral questions is important and is covered on our website.

This book is divided into multiple sections covering the following topics: RTL Verilog coding syntax, RTL Logic Design (including low power RTL design principles), clocking and reset circuits, clock domain crossing questions, digital design fundamentals, and logical thinking questions. Each section is unrelated to the other so you can jump around to any section or question that interests you.

This book is a great starting place for you to begin preparing for your job interview. This book provides you with a broad range of information and covers many topics. By the end of this book, you will have more knowledge and insight into the types of digital design interview questions being asked in the field of semiconductor digital design.

Remember that life will always bring about change, and it's how well you can transition and adapt that is important. Have a strong and positive attitude and you will succeed!

**Good luck on your new journey!**

# RTL Verilog Syntax Questions

# 1. What is the difference between blocking and non-blocking statements, and when are they used?

Blocking statements are coded in Verilog with the = operator, and are used when creating combinatorial logic. This operator *blocks* the simulator from executing subsequent statements until the current evaluation and assignment is done. Consider the following code (E is assigned the immediate new value of C):

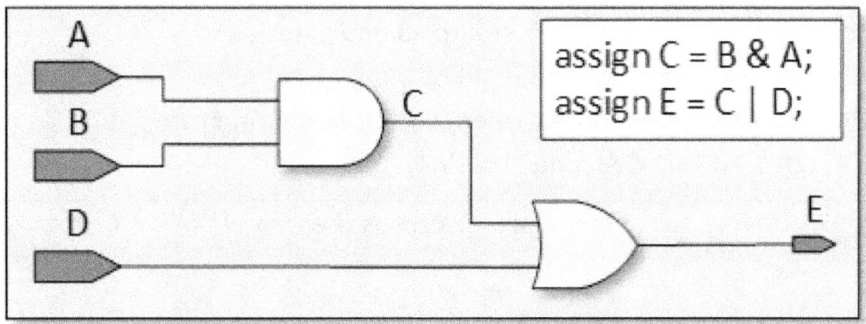

**Blocking Statements and Equivalent Gates** *(Fig. 1)*

Non-blocking statements are coded in Verilog with <= operator, are always used when coding flip flops inside a clocked process. The assignment is postponed until all the subsequent statements are evaluated. This allows for parallel or concurrent execution of statements. In this example, E is assigned the previous value of C (not the immediate):

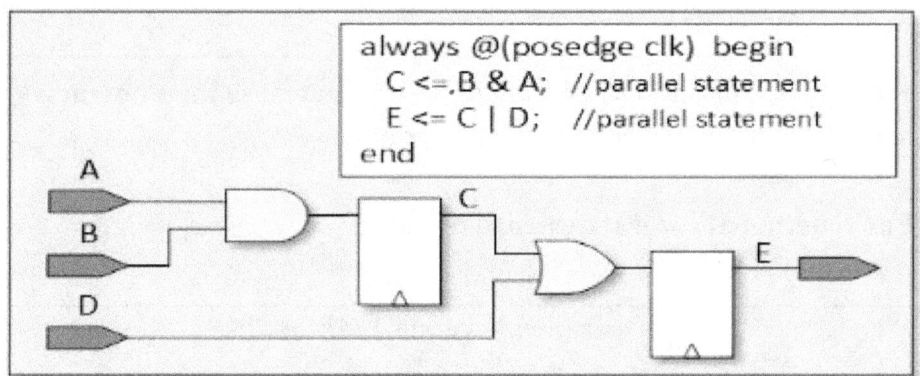

**Non-blocking Statements and Equivalent Gates** *(Fig. 2)*

## 2. Explain the difference between logical and bitwise operators

In Verilog, you should understand the different syntax used for writing conditional if statement operations compared to writing code for creating logical gates with bitwise operations.

---

**For bitwise operations resulting in an *AND* gate, use the & operator**
assign C = A & B;  //This will create an *and* gate

**For conditional *and* if statements, use the && operator**
if ( cond1 == 1'b1 && cond2 == 1'b1) {...}

**Operator:  & versus &&** *(Ex. 1)*

---

**For bitwise operations resulting in an *OR* gate, use the | operator**
assign C = A | B;  //This will create an *or* gate

**For conditional *or* if statements , use the || operator**
if ( cond1 == 1'b1 || cond2 == 1'b1) {...}

**Operator:  | versus ||** *(Ex.2 )*

---

**For bitwise operations resulting in *inverter* gate, use the ~ operator**
assign A = ~B;  //This will create an *inverter*

**For conditional *not* if statements, use the !**
if (!cond1) { ... }

**Operator:  ~ versus !** *(Ex. 3)*

---

# 3. Write code for logic gates: *and, or, xor, nand, nor, xnor*

```
assign C =  (A & B);    //AND gate
assign C =  (A | B);    //OR gate
assign C =  (A ^ B);    //XOR gate
assign C = ~(A & B);    //NAND gate
assign C = ~(A | B);    //NOR gate
assign C = ~(A ^ B);    //XNOR gate
```

**Bitwise Operators** *(Ex. 4)*

# 4. How can you bitwise reduce a multibit signal?

```
wire [15:0] databus;
wire all_ones_detected;
wire is_databus_odd;
wire signal_not_zero;

assign all_ones_detected = &databus; // AND all the bits together;
assign is_databus_odd    = ^databus;  // XOR all the bits together;
assign signal_not_zero   = |databus; //   OR all the bits together;
```

**Bitwise Reduction** *(Ex. 5)*

# 5. What code multiplies or divides by powers of 2?

```
assign C = A << 2;  // shift left ( multiply by 4)
assign C = A >> 3;  // shift right (division by 8)
```

**Shift Operations** *(Ex. 6)*

# 6. How would you perform sign extension?

```
reg [4:0] A;
wire [9:0] C;
assign C = { {5{A[4]}} ,A}; // sign extension
```

**Concatenation Operations** *(Ex. 7)*

# 7. What are three ways to code a 4:1 mux in Verilog?

Assume the 4:1 mux inputs are named 'A, B, C, D', you will need a 2-bit *select* line (to choose between 4 inputs) and have a 1-bit *output*:

```
Coding Style 1:
assign output = (select == 2'b00 ) ? A :
                (select == 2'b01) ? B :
                (select ==2'b10) ? C : D;


Coding Style 2:
always @(*)
if (select ==  2'b00)
    output <= A;
else if (select == 2'b01)
    output <= B;
else if (select == 2'b10)
    output <=C;
else
    output <= D;


Coding Style 3:
always @(*)
case (select):
  2'b00: output  <= A;
  2'b01: output  <= B;
  2'b10: output  <= C;
  2'b11: output  <= D;
end case;
```

**Verilog coding styles for a 4:1 mux** *(Ex. 8)*

You could also instantiate a mux cell directly if it exists in your library.

# 8. What circuit would synthesis create for previous mux coding styles?

Coding Styles 1 and 2 are equivalent circuits but use different coding styles, and both produce a priority encoder (path A is highest)

| | |
|---|---|
| **Style 1:**<br><br>assign output = (select == 2'b00) ? A :<br>(select == 2'b01) ? B :<br>(select ==2'b10) ? C :<br>D;<br><br>**Style 2:**<br>always @(*)<br>if (select == 2'b00)<br>    output <= A;<br>else if (select == 2'b01)<br>    output <= B;<br>else if (select == 2'b10)<br>    output <=C;<br>else<br>    output <= D; | 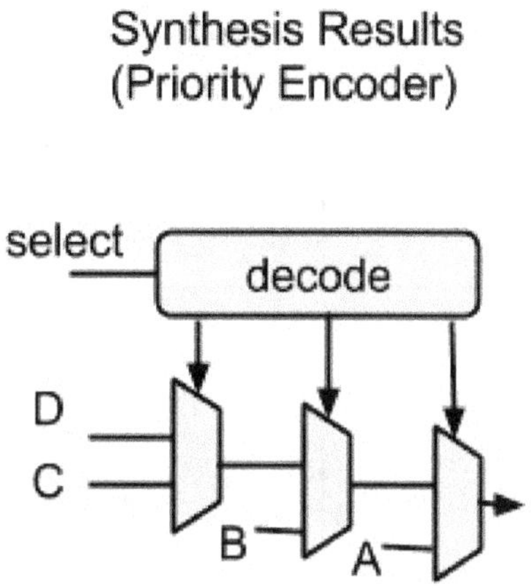 Synthesis Results<br>(Priority Encoder)<br><br>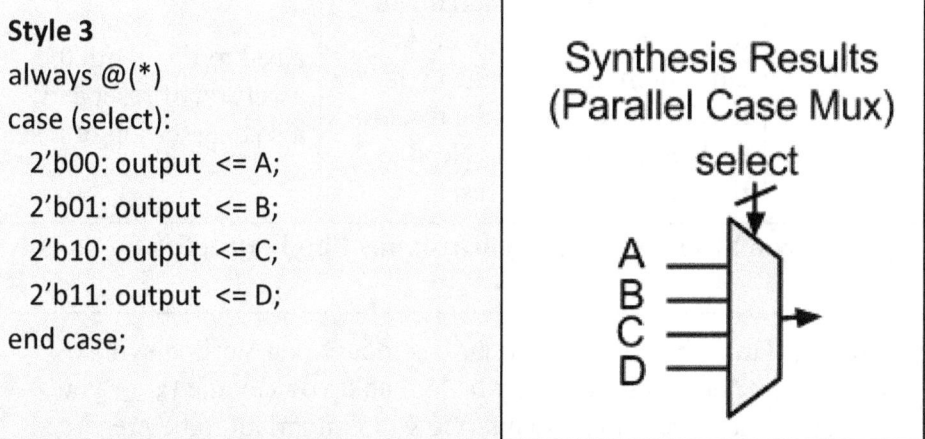 |

**Priority Encoder** *(Ex. 9)*

| | |
|---|---|
| **Style 3**<br>always @(*)<br>case (select):<br>  2'b00: output  <= A;<br>  2'b01: output  <= B;<br>  2'b10: output  <= C;<br>  2'b11: output  <= D;<br>end case; | Synthesis Results<br>(Parallel Case Mux)<br>select<br><br>A<br>B<br>C<br>D |

**Parallel Muxing Scheme using full case statement** *(Ex. 10)*

# 9. Write Code for Asynchronous/Synchronous Flip Flops and discuss the pros and cons of each

| | Asynchronous | Synchronous |
|---|---|---|
| Verilog Code | always @(posedge clk or negedge rst)<br>　If (!rst )<br>　　Q <= #1 1'b0;<br>　else<br>　　Q <= #1 D; | always @(posedge clk)<br>　If (!rst )<br>　　Q <= #1 1'b0;<br>　else<br>　　Q <= #1 D; |
| Schematic | | |
| Pros | No need for clocks to be running | No need to worry about asynchronous timing |
| Cons | Any glitch on reset signal will reset flops (If there are cross clock domain paths, then you need to synchronize the reset to each clock domain). Need to time the removal of reset | Need to time both the assertion of reset and also deassertion of reset |

**Asynchronous versus Synchronous Flip Flops** *(Table 1)*

**Author's Tip**: The questions so far has focused on Verilog syntax. Basic syntax questions can easily be looked up on Google (<u>after</u> you have a job), however, on a job interview it's important to be prepared for these questions and knock them out of the park.

# 10. Write Verilog which captures the input signal below to both a register and a latch. Draw the outputs of each.

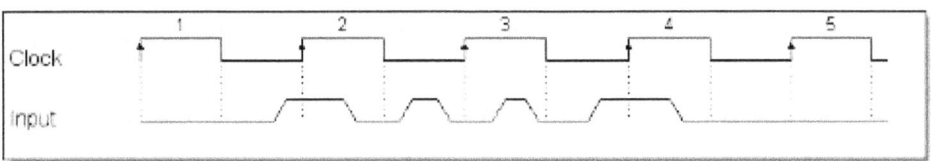

**Timing Diagram - capture input into a latch and flip flop** *(Fig. 3)*

```
//purposely coding latch
always @(clk or input)
if (clk == 1'b1)
   Q_latch <= input;

//flip flop coding a Latch
always @(posedge clk)
   Q_flop <= #1 input;
```

**Verilog code for Latch and Flip Flop** *(Ex. 11)*

The outputs of the latch and flip flop is below in the timing diagram:

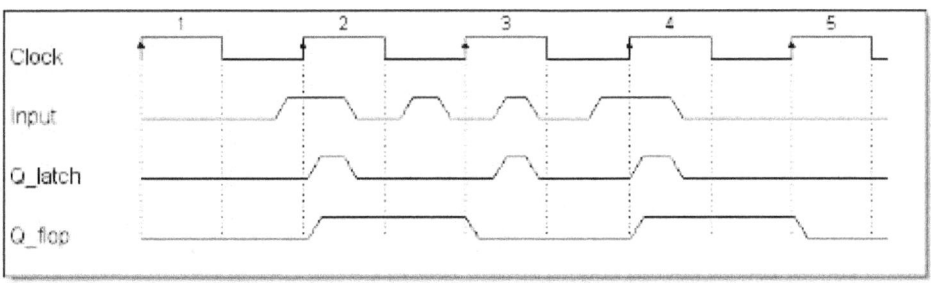

**Latch vs. Flip Flop timing diagram** *(Fig. 4)*

The output of the flip flop only changes on the rising edge of clock, and will equal the value of the input captured at the rising edge of clock. The latch output will follow the input signal while the latch is open (in this case while clock is high). When the latch closes (clock is low), the output holds its previous value.

# RTL Logic Design Questions

# 11. Design a circuit that can detect if an input signal transitions from high to low, or low to high.

As with all interview questions, you should state your assumptions before answering the questions and confirm the question with the interviewer so you are both on the same page   We are going to assume the input signal is driven from the same clock domain as our detection circuit.

The easiest way to detect if an input signal has changed is to simply compare the currently sampled signal at time T to the previous version of the signal 1 clock period before (at time T-1).  Therefore,  a flip flop is used to capture and delay the signal, and then you can compare it to the previous version of the same signal:

```
//capture the signal to have a 1-clock delayed version
always @ ( posedge clk or negedge reset)
if (~reset)
  Q <= #1 1'b0;
else
  Q <= #1 D;

//detect signal transitions
assign Toggle = (Q ^ D) ; //xor gate
assign Falling = ~D & Q; // D is low, but was high
assign Rising  =  D & ~Q; // D is high, but was low
```

**Verilog Code for Edge Detection** *(Ex. 12)*

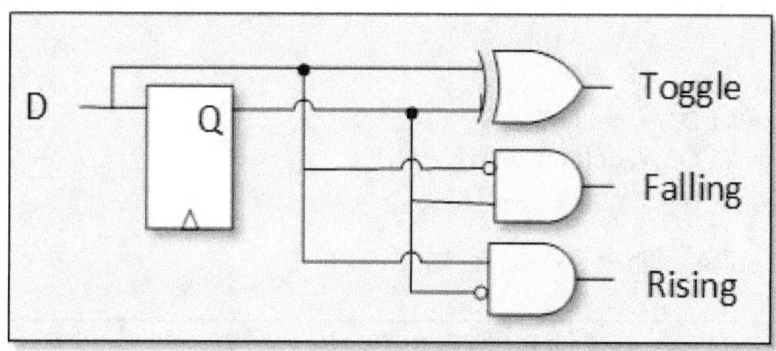

**Equivalent Gates from RTL code in Ex. 11** *(Fig. 5)*

The timing diagram is shown below (drawn with free TimeGen software):

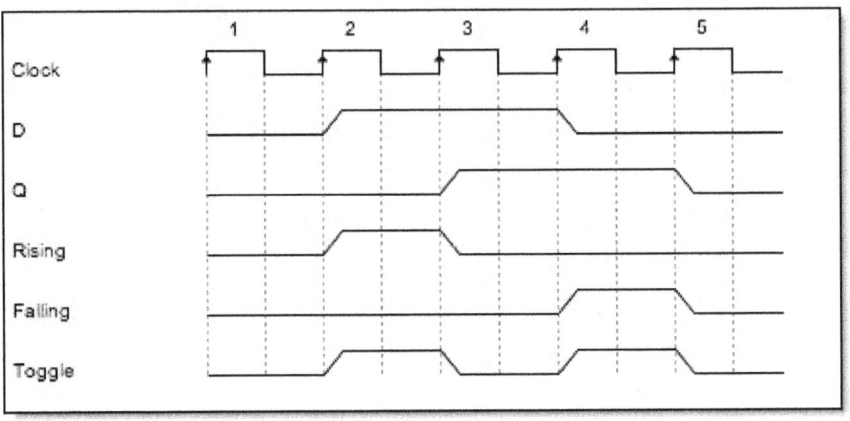

**Timing Diagram for Edge Detection using clocks** *(Fig. 6)*

# 12. Design a circuit to detect a 1 cycle high pulse on an input (detect an input signal transition from 0->1->0)

Using delay flip flops to hold the state of the signal for the previous two cycles, you can simple check for 0, 1, then 0. You could also use an FSM.

```
//capture the signal and delay 2 cycles

always @ ( posedge clk or negedge reset)
if (~reset) begin
  Q1 <= #1 1'b0;
  Q2 <= #1 1'b0;
end
  else begin
  Q1 <= #1 D;  //delay T-1
  Q2 <= #1 Q1; //delay T-2
  end;

assign pulse_high= ~D & Q1 & ~Q2;
```

**Circuit to detect pulse** *(Ex. 13)*

# 13. Design a sequence detector for the pattern: 1,0,1,1,0

One method is to design a FSM to detect the sequence. If the FSM reaches state 5, then generate a valid output pulse (all the other states will output low). It's also important to make sure you go back to correct state if the wrong value comes in (you don't always need to go back to idle state). You need to check if some of the sequence has already started and go back to the appropriate state.

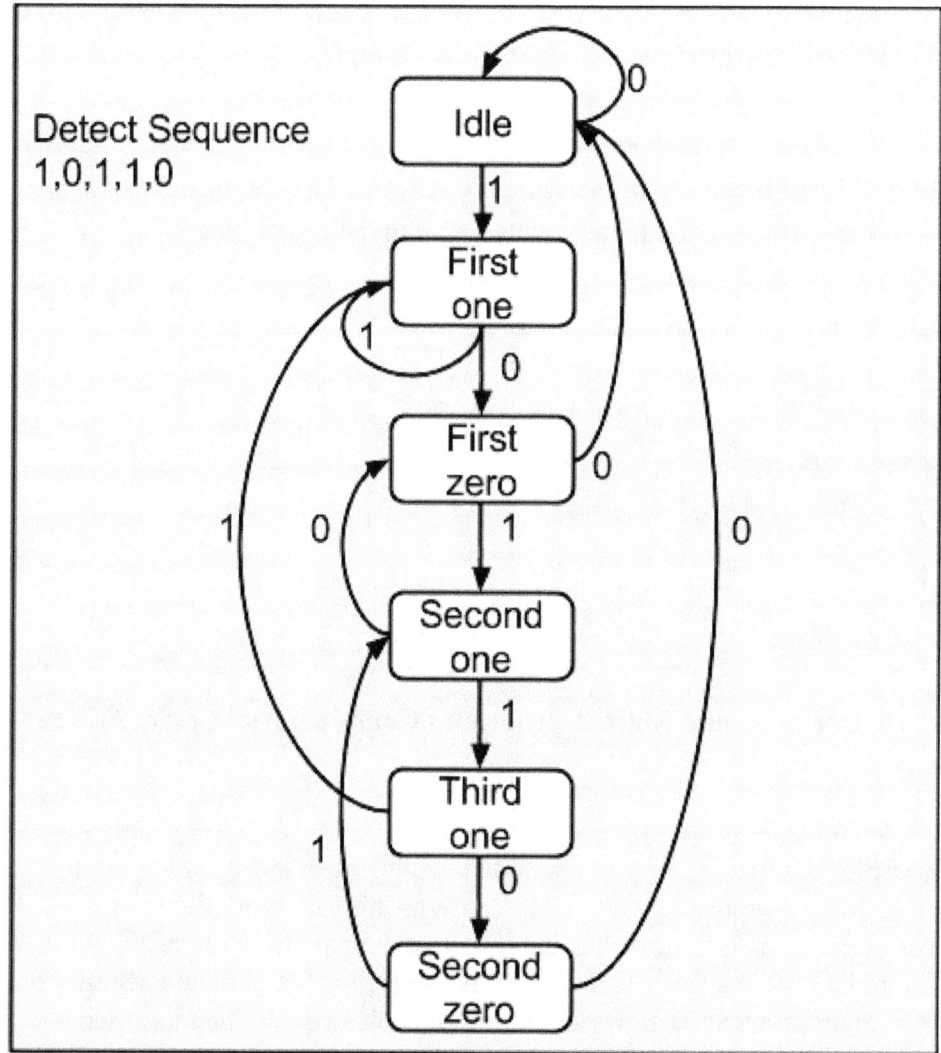

**Sequence Detector using FSM** *(Fig. 7)*

# 14. Detect if pattern 10110 appears anywhere in the last 5 inputs

To detect a pattern, we just need to design a decoder circuit. You could use shift registers and then add combinatorial logic to decode the register outputs to check for a match. Whenever the values match, a pulse is generated. Be sure to clarify with the interviewer if the circuit needs to start over if the sequence is wrong (or if they want you to just detect anytime the pattern appears). If you need to detect the sequence, then you will need a state machine in the previous example.

The below method requires N number of shift registers to hold the length (N) of the string or the sample window. This circuit will always output true if the string matches anywhere in the last N inputs.

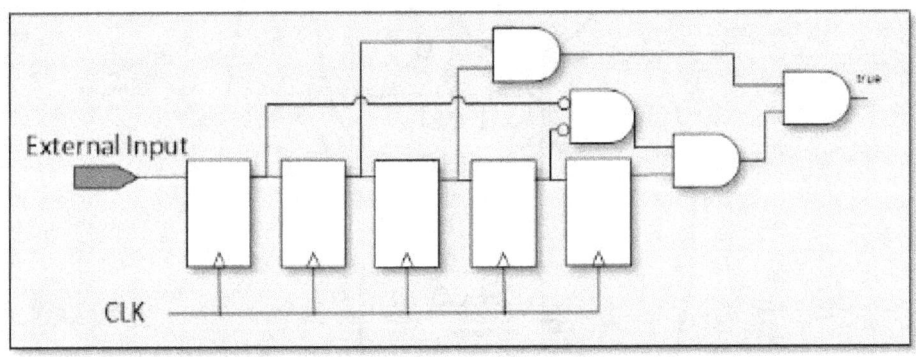

**Decoder Circuit - Shift Register and Combinatorial Logic** *(Fig. 8)*

**Author's Tip:** Be sure to understand if the interview question requires a FSM or a simple decode. Also, how would you work the above example if the question changed to detect the pattern in the last 8 samples? The answer is to use a 8 bit shift register, and then decode for the appropriate sequence after output of each stage. Then you will need to logically or the outputs of each stage

# 15. Given the timing diagram, write the Verilog code to produce the Start signal and the chip selects:

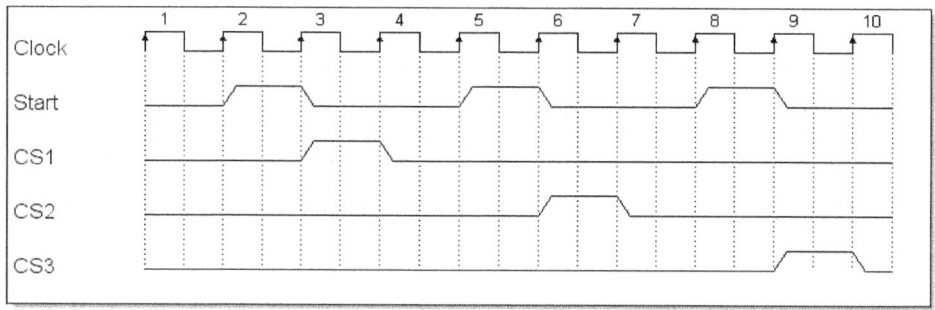

**Write the Verilog code to produce the above waveform** *(Fig. 9)*

```
//generate a counter to 3 (used to create start pulse)
always @(posedge clk or negedge rst)
  if (!rst)
    counter <= 2'b00;
  else if (counter == 2'b10)
    counter <= 2'b00;
  else
    counter <= counter + 1'b1;

assign start = (counter == 2'b01) ? 1'b1 : 1'b0; // start pulse

always @(posedge clk or negedge rst)
  if (!rst)
    cs <= 3'b000;
  else if (start == 1'b1)
  begin
    case (cs):
    3'b000:  cs <= 3'b001;  //instead of a case statement, we could use shift
    3'b001:  cs <= 3'b010;  //registers
    3'b010:  cs <= 3'b100;
    3'b100:  cs <= 3'b000;
    default:  cs <= 3'b000;
    endcase
  end

assign CS1 = cs[0];  assign CS2 = cs[1];  assign CS3 = cs[2];
```

**Example Verilog Code for above waveform** *(Ex. 14)*

# 16. Design a debounce circuit which removes glitches, with the following assumptions:

1) The circuit should filter out any noise or glitches on the input signal. The input signal must be valid high for at least 2 clocks
2) Assume the input signal is driven from an asynchronous clock.
3) Generate a pulse when the signal transitions from low-to-high.

Since the question stated to assume the external input signal is driven from an asynchronous clock, the circuit we need to design must use a synchronizer on the input.

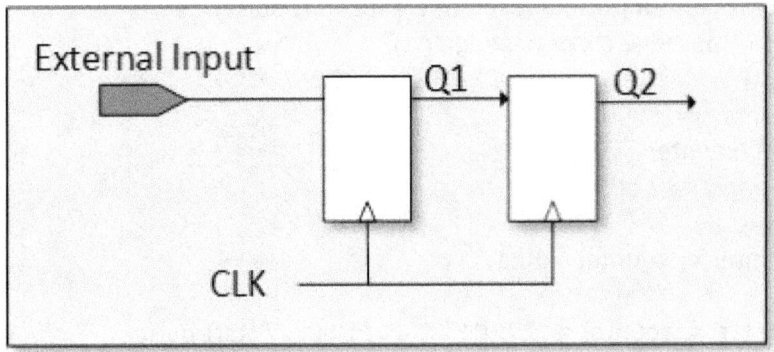

**Classic Synchronizer Circuit using 2 Flip Flops** *(Fig. 10)*

After the synchronizers, the signal must be delayed two more cycles so we can check that is is still valid, and to also detect a rising edge.

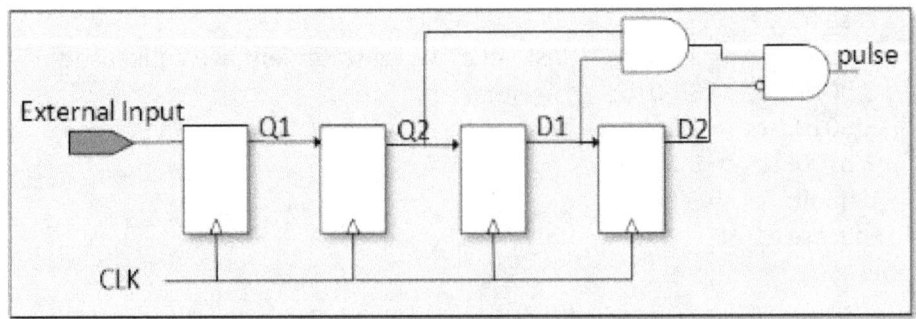

**Circuit to detect low to high transition** *(Fig. 11)*

# 17. Write Verilog Code to generate a Gray Code counter

A valid question for a digital designer is to write RTL code to generate gray code. I've seen two ways to accomplish this. The first method is straightforward code without any tricks. The second method is a quicker method and something you should be aware to translate binary counter to gray code counter.

Below is RTL code written to generate a 3-bit gray code counter:

```
always @(posedge clk or negedge rst)
  if (!rst)
    counter <= #1 3'b000;
  else
    counter <= #1 counter + 1;
end

always @(counter)
    case (counter):
      3'b000 :   gray_output <= 3'b000;
      3'b001 :   gray_output <= 3'b001;
      3'b010 :   gray_output <= 3'b011;
      3'b011 :   gray_output <= 3'b010;
      3'b100 :   gray_output <= 3'b110;
      3'b101 :   gray_output <= 3'b111;
      3'b110 :   gray_output <= 3'b101;
      3'b111 :   gray_output <= 3'b100;
      default:   gray_output <= 3'b000;
    end case;
```

**Verilog Code for 3-bit Gray Code** *(Ex. 15)*

One logical solution would be to create a 3-bit register and use it as binary counter that is incremented by 1 each clock cycle. A signal named "gray_output" decodes the binary counter and then generates the new gray code output. The problem with this approach is for a N-bit counter you need to explicitly decode each state.   A better, more generic solution using an XOR gate is presented below:

| BCD | Gray Code | |
|---|---|---|
| 0000 | 0000 | |
| 0001 | 0001 | |
| 0010 | 0011 | |
| 0011 | 0010 | |
| 0100 | 0110 | |
| 0101 | 0111 | |
| 0110 | 0101 | |
| 0111 | 0100 | |
| 1000 | 1100 | |
| ..... | .... | |
| 1111 | 1000 | |

**BCD to Gray Code** *(Fig. 12)*

```
always @(posedge clk or negedge rst)
  if (!rst)
    bcd_counter <= #1 4'b0000;
  else
    bcd_counter <= #1 bcd_counter + 1; // binary counter
end

//convert binary to gray code
assign gray_code = { bcd_counter[3] ,
              bcd_counter[3] ^ bcd_counter[2],
              bcd_counter[2] ^ bcd_counter[1],
              bcd_counter[1] ^ bcd_counter[0]}
```

**Verilog code to convert BCD to Gray Code** *(Ex. 16)*

**Author's Tip**: I was asked some variation of a gray coding question on 80% of my interviews.  It is important in today's low power design ( since only 1 bit toggles saving dynamic power).  Also, gray coding can be used to send a multi-bit vector across clock domains (after classic double synchronization).

# 18. Design a synchronous FIFO module and logic that pushes and pops data using a dual port RAM. The FIFO logic should generate a fifo full and a fifo empty signal.

---

**Dual Port RAM datasheet**
**Inputs/Outputs:** (inputs clk, rst, write, write_address, write_data, read_address; output read_data);
**Description:** : A dual port RAM allows both writes and reads to occur simultaneously by using two separate write and read address ports.
**Writes:** When the *write* signal is asserted high, data on the write_data bus is written into location pointed to by write_adresss.
**Reads:** The read address points to a location to retrieve the data on put onto the read_data bus.
**Width x Depth:** The RAM can store 256 entries

.

The code is listed below which generates the FIFO control logic:

```
module FIFO
input wire clk, rst, push, data_in, pop, data_in;
output fifio_full, fifo_empty, data_out);

reg[7:0] write_address;
reg[7:0] read_address;
reg[7:0] fifo_count;

//generate internal write address
always @(posedge clk or negedge rst)
if (!rst)
  write_address <= #1 8'b0000_0000;  //256 locations
else if (push == 1'b1)
  write_address <= #1 write_address +1'b1;

//generate internal read address pointer
always @(posedge clk or negedge rst)
if (!rst)
  read_address <= #1 8'b0000_0000;  //256 locations
```

```
else if (pop == 1'b1)
  read_address <= #1 read_address +1'b1;

//generate FIFO count
// increment on push, decrement on pop
always @(posedge clk or negedge rst)
if (!rst)
  fifo_count <= #1 8'b0000_0000;  //256 locations
else if (push == 1'b1 && pop == 1'b0)
  fifo_count <= #1 ffio_count +1;
else if (push == 1'b0 && pop == 1'b1)
  fifo_count <= #1 ffio_count - 1;

//generate FIFO signals
assign fifo_full    = (fifo_count == 8'b1111_1111) ? 1'b1 : 1'b0;
assign fifo_empty = (fifo_count == 8'b1111_1111) ? 1'b1 : 1'b0;

//Connect RAM
i_ram RAM ( .clk (clk), .rst(rst), .write(push),
.write_address(write_address), .write_data(data_in), .read_address
(read_address), .read_data(data_out));
```

**FIFO control logic** *(Ex. 17)*

**Author's Tip:** Note this solution presented uses a fifo count register (which adds another 8 registers) . Another method which saves area is to extend the write and read address pointers by 1 extra bit. When the pointers are equal, the fifo is empty. When the MSBs are different, but remaining bits[7:0] are equal, the fifo is full.

Also note that if this question used asynchronous clock domains, the circuit would need to synchronize any signals that cross domains such as fifo full or empty. But when you pass these signals across domains, the synchronization possibly causes an extra delay (due to the sync flop) which would could lead to overflows and underflows. To solve this, you can use gray coded write and read address pointers. The gray coded address pointers are then synchronized into the opposite clock domains, and then synchronously generates fifo_full and fifo_empty.

# 19. Design a circuit to detect if a binary number is divisible by 3. At reset, the binary number is 0. On each clock cycle, a new binary digit will be shifted in 1-bit at a time (into the LSB location). If the new binary value is divisible by 3, output a 1.

The trick to this is to divide the binary number by 3 and then check the remainder (modulo value). The maximum number of states we could have based on the modulo results is just 3 (mod 0, mod 1, or mod 2).

Our FSM will start in state 0 which represents that the number is divisible by 3. If the module is 1, then we go to state 1. If modulo is 0, we go back to state 0 (and generate div_by_3 high). Each state represents the module value. For this example, let's assume the first input bit is 0, then 1, 1,0,1,0,0,1:

| binary num | Div by 3? | State |
|---|---|---|
| 0 | yes | mod 0 |
| 01 | no | mod 1 |
| 011 | yes | mod 0 |
| 0110 | yes | mod 0 |
| 01101 | no | mod 1 |
| 011010 | no | mod 2 |
| 0110100 | no | mod 1 |
| 01101001 | yes | mod 0 |

assign div_by_3 = (fsm == mod_0) ? 1'b1: 1'b0;

**FSM states for divisible by 3 circuit** *(Ex. 18)*

**Author's Tip:** You should practice other divisible values such as 4 or 5

## 20. Design a circuit to generate the Fibonacci series. The circuit should have an enable signal which allows the circuit to operate. When enable is low, the circuit should not advance to the next number.

By definition, the first two numbers in the Fibonacci sequence are 0 and 1, and the next number in the sequence is calculated by adding together the previous two numbers. For example: 0, 1, 1, 2, 3, 5, 8, 13,

Since the circuit must self generate the sequence, the design must have at least 2 registers to hold the starting numbers of 0 and 1.

```verilog
//Define registers and signals
input wire clk, rst, enable;
output wire [31:0] sum;

reg [31:0] cur_num;
reg [31:0] next_num;

always @(posedge clk or negedge rst)
  if (!rst)
    cur_num <= #1 32'b0;  //initialize to 0
  else if (enable == 1'b1)
    cur_num <= #1 next_num ;
end

always @(posedge clk or negedge rst)
  if (!rst)
    next_num <= #1 32'b1;  //initialize to 1
  else if (enable == 1'b1)
    next_num <= #1 sum;    //save the sum
end

//The output is the sum of the two numbers
assign sum = cur_num + next_num ;
```

**Verilog code for Fibonacci Generation** *(Ex. 19)*

Below is the equivalent block level diagram that illustrates what the previous Verilog code will create:

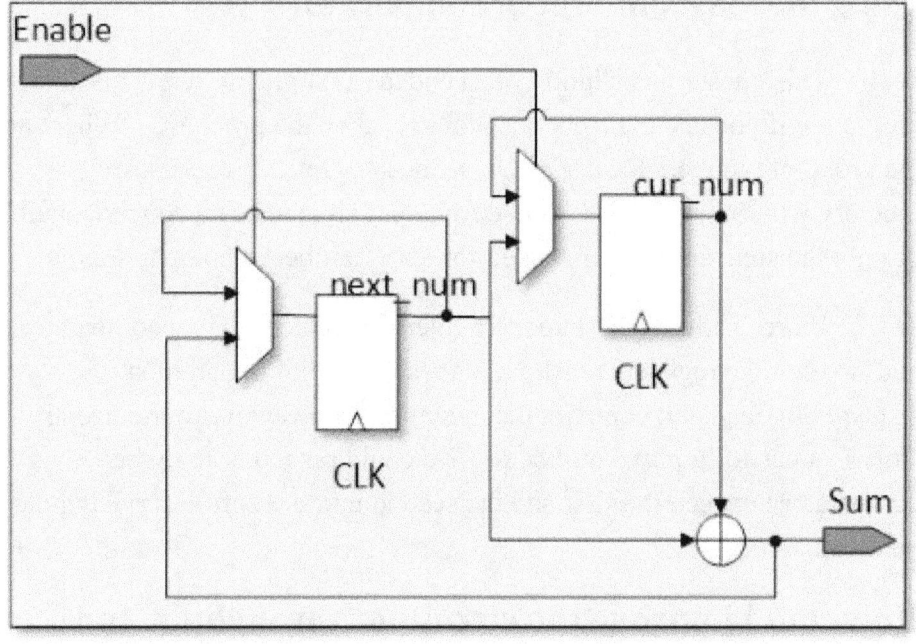

**Circuit to Generate Fibonacci Series** *(Fig. 13)*

For further preparation, you should try to convert other mathematical or computer science algorithms from software to a hardware digitally equivalent circuit.

---

**Author's Tip:**  You should also understand the different tradeoffs and relationships between power, performance, and area when making decisions about the architecture of a digital circuit.  That is, understanding the concept of getting better performance (by using parallel hardware which uses more area) compared to using least amount of gates (resource sharing) and doing things slower over more clock cycles.

---

## 21. Design a circuit to find the maximum value and the second maximum value from a set of binary numbers, using the least amount of comparators.

This question is slightly open ended (as some interview questions are) to see if you ask the right questions to solve the problem. Where are the group of numbers located? Do we need to fetch them first from memory with burst accesses? Is throughput and latency a concern at all, or only the area since we are asked for least number of comparators?

If area is the highest concern, then all we need is one comparator and two storage registers (each with a reset value of 0) and then sequentially read and compare the incoming value against our currently stored values for top two numbers. We could resource share the comparator between the first and the second number before servicing the next value.

## How could improve the circuit for throughput and make it faster? Assume area is of the least importance.

A faster method would be do all the comparator operations in parallel using a binary tree, and use pipeline stages for faster bandwidth.

You could compare the first and second number and the winner would advance. In parallel the third and fourth numbers are compared and winner advances. Using this approach the maximum number will continue to the final pipeline stage. In parallel to the winner's bracket, you would need to also have a loser's bracket in order to find the 2nd largest number, but the same binary tree approach is applied.

> **Arthur's Tip:** This follow up question really is testing if you understand the tradeoffs between performance and area (doing operations in parallel and pipelining) as opposed to doing it serially as the first question (less area but lowest bandwidth).

## 22. Design a circuit that will output 3 signals: *second, hour,* and *minute.* The circuit will receive a 1-bit synchronous input pulse which occurs every 1 ms.

```
//Define inputs
input clk, 1_ms_pulse;
output second, minute, hour;

wire second, minute, hour;
reg [9:0] ms_counter;  //10 bit counter
reg [5:0] second_counter;  //6 bit counter
reg [5:0] minute_counter;  //6 bit counter

always @(posedge clk or negedge rst) begin
  if (!rst)
    ms_counter <= #1 10'b0;  //initialize to 0
  else if (1_ms_pulse == 1'b1) begin
    if (ms_counter == 999)
      ms_counter <= #1 10'b0;  //reset to 0
    else
      ms_counter <= #1 ms_counter + 1'b1 ;
  end

always @(posedge clk or negedge rst)
  if (!rst)
    second_counter <= #1 6'b0;  //initialize to 0
  else if (second == 1'b1) begin
    if (second_counter == 59)
      second_counter <= #1 6'b0;  //reset to 0
    else
      second_counter <= #1 second_counter + 1'b1 ;
  end

always @(posedge clk or negedge rst)
  if (!rst)
    minute_counter <= #1 6'b0;  //initialize to 0
  else if (minute == 1'b1) begin
    if (minute_counter == 59)
```

*(cont. on next page)*

```
    minute_counter <= #1 6'b0;  //reset to 0              (continued   )
  else
    minute_counter <= #1 minute_counter + 1'b1 ;
  end

assign second = (ms_counter == `d999 && 1_ms_pulse) : 1'b1 : 1'b0;
assign minute = (second_counter == `d59 && second)  : 1'b1 : 1'b0;
assign hour   = (minute_counter == `d59 && minute)  : 1'b1 : 1'b0;
```

**Verilog code to generate one second, minute, and hour** *(Ex. 20)*

# 23. The below diagram has 2 input signals: Clock and A. Write Verilog code to produce the output signal B.

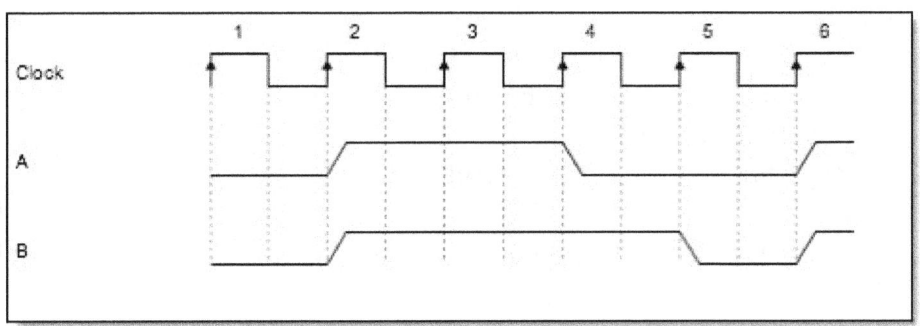

**Write Verilog Code to Produce this Timing Diagram** *(Fig. 14)*

Judging from the output waveform B,  when the input A when it is high the output follows it, but when the input falls the output is delayed by 1 cycle.   The following code will produce the above timing diagram:

```
//flip flop to delay A input
always @(posedge clk or negedge rst)
  if (!rst)
    A_delay <= #1 1'b0;
  else
    A_delay <= #1 A;
assign B = A | A_delay ;  //or gate
```

**Verilog Code for Above Timing Diagram** *(Ex. 21)*

## 24. Draw the structure of a digital 5-TAP FIR Filter

   Digital filters can be used in various application such as audio or video processing.   There are two types of digital filters:  FIR and IIR. The FIR (Finite Impulse Response) circuit does not have a feedback loop. IIR (Infinite Impulse Response) circuits feedback the output to the intput. The filter consists of multiply, add, and accumulate (MAC) function which creates the output.   Each register below represents a tap, with each tap multiplied by a coefficient and then summed together for the output.

   In video processing, upscaling an image (zooming in), requires shifting in 1 bit input value but then outputting 2 values.  On the first cycle, one set of coefficients are loaded and then a MAC function is used to create an output value.   On the next cycle, a 2nd set of coefficients are loaded and the MAC function generates a 2nd output value.  The process repeats itself and the pipeline moves forward.  The opposite is true for downscaling, in which the input rate is twice as fast as the output rate (shift in 2 values, but only output 1 value).   This is a simplified explanation but it gives you an idea of how filters are used.

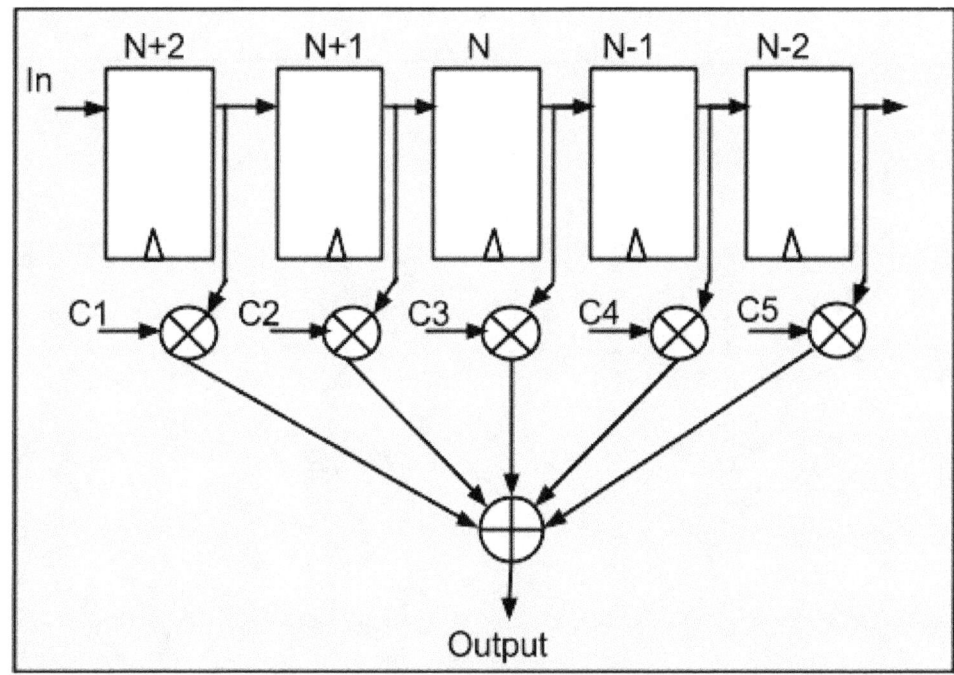

**5-TAP Digital FIR filter** *(Fig. 15)*

# Clock Divider, Clock Gating, and Reset Questions

# 25. Write Verilog code for clock divide by 2 circuit

As a digital design engineer, you should understand how to code and design any clock divider circuit (both even or odd), and make sure it is glitch free. Below is an example of simple divide by 2 circuit:

```
//divide by 2 clock (toggle)
always @(posedge clk or negedge rst)
if (!rst)
   clk_div_2 <= #1 1'b0;
else
   clk_div_2 <= #1 ~(clk_div_2);
```

**Verilog Code for Clock Divide by 2** *(Ex. 22)*

# 26. Clock Divide by 3 circuit with 50-50 duty cycle

For generating 50-50 duty cycle clock output for odd dividers, you will need to use both the *rising* edge and *falling* edges of input clock to produce the final generated output clock. Also, to guarantee glitch-free logic, the logic that drives the final output clock should guarantee (by design) that only 1 input signal to the combinatorial logic is changing at a time, and that input should be driven from a flip flop. Otherwise, you will create glitchy logic due to race conditions.

```
always @(posedge clk or negedge rst)
  if (!rst)
     posedge_cnt  <= #1 2'b00;
  else if (posedge_cnt == 2'b10)
     posedge_cnt  <= #1 2'b00;
  else
     posedge_cnt <= #1 posedge_cnt + 1;
```

*(cont. on next page)*

```
always @(posedge clk or negedge rst)                    (continued)
  if (!rst)
     rise_pulse_reg <= 1'b0;
  else if (posedge_cnt == 2'b01)
     rise_pulse_reg <= #1 1'b1;
  else
    rise_pulse_reg <= #1 1'b0;

always @(negedge clk or negedge rst)
  If (!rst)
     neg_pulse_reg <= #1 1'b0;
  else
     neg_pulse_reg <= #1 rise_pulse_reg;

assign clk_output = rise_pulse_reg | neg_pulse_reg; //glitch free
```

**Verilog Code for Clock Divide by 3 with 50-50 Duty Cycle** *(Ex. 23)*

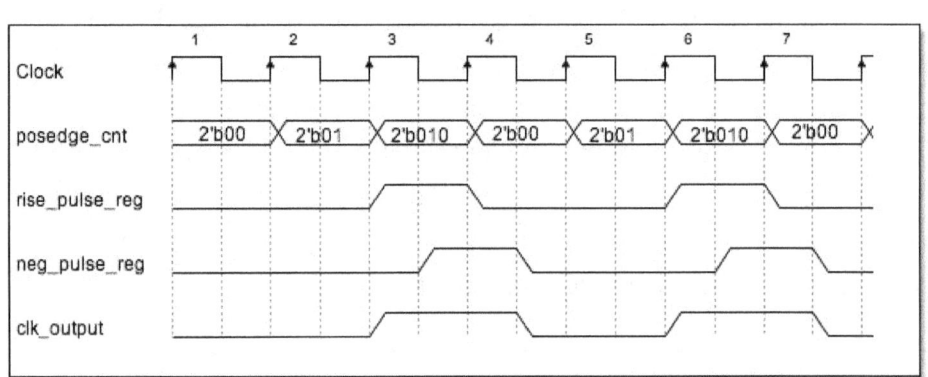

**Timing Diagram for Clock Divide by 3** *(Fig. 16)*

**Author's Tip**: For clock dividers with an *even divisor* value, create a counter or FSM to toggle the output clock when the counter is half way. For *odd divider* values when duty cycle does not matter, just have one extra state where the clock is held high. For odd number divisors that require a 50-50 duty cycle, be sure the output is *glitch free*.

# 27. Write Verilog for a generic clock divide by N circuit.

```verilog
reg [7:0] divider_value;           //programmed to value minus 1

always @(posedge clk or negedge rst)
  if (!rst)
    posedge_cnt  <= #1 {8{1'b0}};
  else if (posedge_cnt == divider_value)
    posedge_cnt  <= #1 {8{1'b0}};  //reached end, start count over
  else
    posedge_cnt <= #1 posedge_cnt + 1;

always @(posedge clk or negedge rst)
  if (!rst)
    rise_pulse_reg <= #1 1'b0;
  else if (posedge_cnt == divider_value[7:1])
    rise_pulse_reg  <= #1 1'b1; //set high (half way)
  else if (posedge_cnt == divider_value)
    rise_pulse_reg  <= #1 1'b0; //set low (reached end)

//delay by half cycle (only for odd)
always @(negedge clk or negedge rst)
  If (!rst)
    neg_pulse_reg  <= #1 1'b0;
  else if (divider_value[0] == 1'b0) // odd (programmed value minus 1)
    neg_pulse_reg <= #1  rise_pulse_reg;

assign clk_output = rise_pulse_reg |  neg_pulse_reg;
```

**Example Verilog Code for Generic Clock Divide by N** *(Ex. 24)*

There waveforms on the next page show clock divider values for divide by 3, divide by 4, and divide by 5.

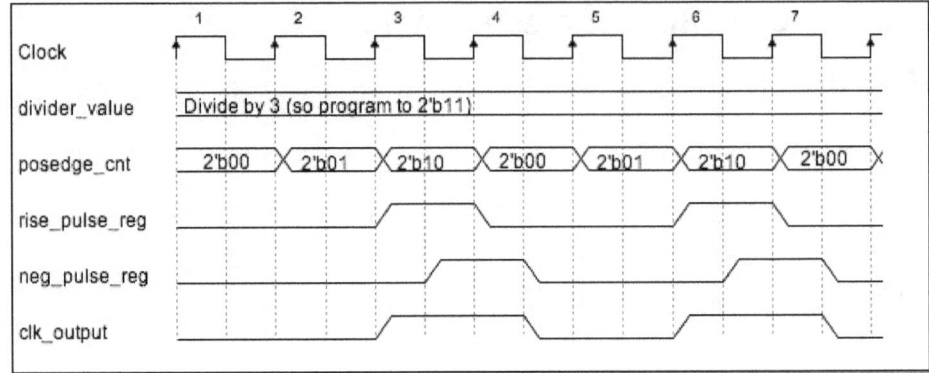

**Timing Diagram for Clock Divide by 3 with (50-50 duty cycle)** *(Fig. 17)*

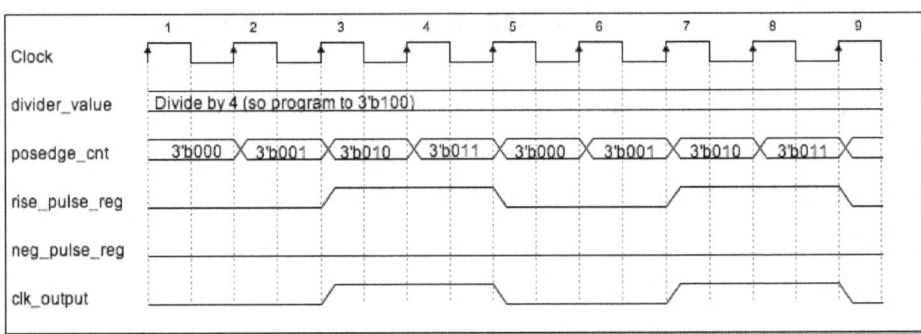

**Timing Diagram for Clock Divide by 4 with (50-50 duty cycle)** *(Fig. 18)*

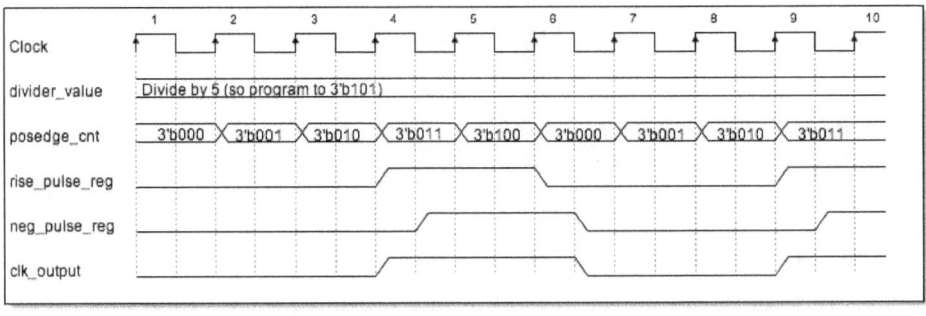

**Timing Diagram for Clock Divide by 5 with (50-50 duty cycle)** *(Fig. 19)*

**Author's Tip**: Remember, it's important to understand how the odd clock divider circuits are glitch free. By our design, we are allowing only 1 input to change to the combinatorial OR gate (and it is driven from a flop), otherwise glitches would occur. You should practice coding and drawing these timing diagrams on your own since these questions are popular on interviews.

# 28. Design a clock gating cell, controlled with an enable signal, that is glitch free

Understanding how to design a glitch free clock gating cell is very important. The below diagram uses a falling edge latch to prevent glitches because the enable signal is not allowed to transition when the clock input is high:

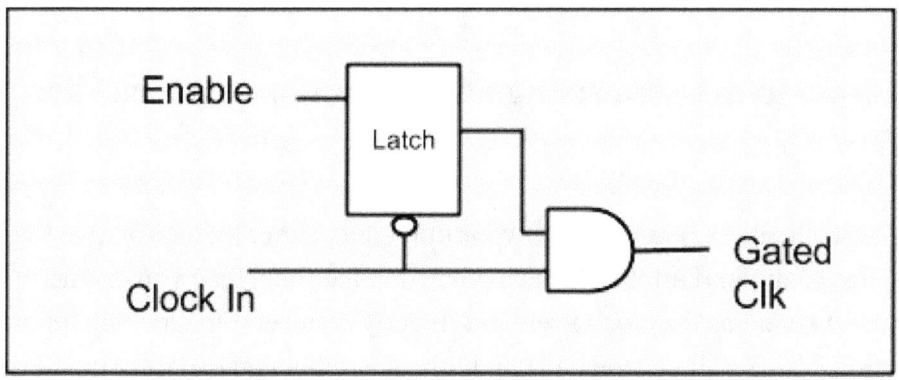

**Latched Based Clock Gating Circuit** *(Fig. 20)*

If the latch was not there, then the enable signal could rise or fall while the clock is high, causing the gated output clock to glitch or have duty cycle less than 50%. The above circuit works because when the input clock rises, the negative edge latch closes and doesn't allow the latch output to move (which keeps the enable signal to the *AND* gate stable even if the input enable signal moves). When the input clock falls, the latch is open and the enable signal is free to come across (since the input clock is low, it's ok now to gate the clock).

```
always @(clk, enable)
begin
  if (!clk)
    latch_q <= enable;
end

assign gated_clk = latch_q & clk_in;
```

**Example Verilog Code for Glitch Free Clock Gate** *(Ex. 25)*

There is an important timing constraint to understand with the above falling edge latch approach. The enable signal is usually launched on a rising clock edge, but the latch uses the falling edge to capture the enable signal therefore creating a true half cycle path. I've seen interview questions where it was asked why if you increase the clock frequency setup violations would appear in the clock gate. The answer is because of the half cycle path that exist between enable and falling edge latch.

## 29. How to detect a rising edge of a signal if clocks are off?

If the main input clock to your module is currently unavailable to use, for example, if all the clocks are gated at the time, then you can use the input signal as the clock itself and directly connect it to clock pin of a flip flop. This will generate a level high signal that you can use to request for clocks. After the local clocks are up and running you still need to clear the flop, and you need to be careful with the timing and treat the output Q as asynchronous (therefore synchronize it before using it).

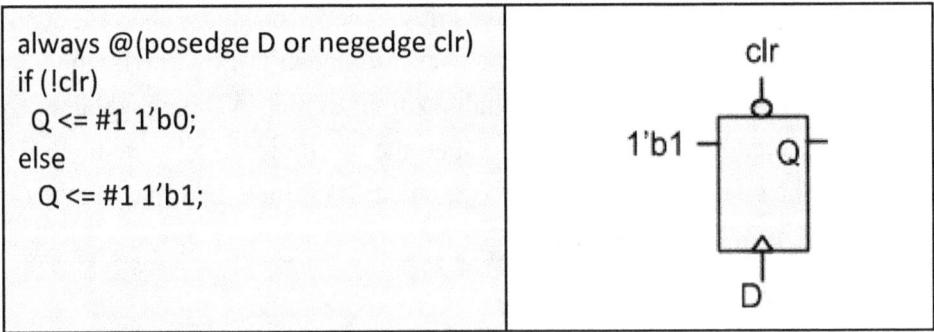

```
always @(posedge D or negedge clr)
if (!clr)
  Q <= #1 1'b0;
else
  Q <= #1 1'b1;
```

**Asynchronous Edge Detection Circuit (when clocks are off)** *(Ex. 26)*

**30. Design a reset synchronizer circuit. Your circuit will receive two input signals: reset, and clock. The circuit should output a new reset signal, which allows asynchronous reset assertions, and synchronized reset de-assertions.**

```
input rst, clk;
output local_reset;

reg reset_deassert_sync;
reg reset_deassert; //register for synchronous deassertion

always @(posedge clk or negedge rst)
if (!rst) begin
  reset_deassert_sync <= # 1'b0;
  reset_deassert      <= # 1'b0;
end
else begin
  reset_deassert_sync <= #1 1'b1;   //1st stage sync (could go metastable)
  reset_deassert      <= #1 reset_deassert_sync; //2nd stage sync flop
end

assign local_reset = rst & reset_deassert;
```

**Verilog code for Reset synchronization** *(Ex. 27)*

# Clock Domain Crossing Questions

# 31. What is metastability?

Metastability occurs when a flip flop has a setup or hold time violation. When either violation occurs, the output of that flip flop becomes metastable, or enters a quasi-stable state , and it may settle either high or low. Metastability can have dangerous consequences if not handled properly in a design. During STA (Static Timing Analysis), all the timing paths inside a circuit will be checked for all setup or hold violations and be fixed. But if the circuit has clock domain crossings (CDC), then the designer must be sure the proper synchronization is implemented so metastability does not occur.

# 32. Design a circuit to synchronize a signal from a slow clock domain to a fast clock domain

Below is a classic clock domain crossing circuit for a 1-bit signal. If the source signal is from a slower clock domain, and if the destination clock domain is at least 2x faster, then you can use a 2-stage synchronizer:

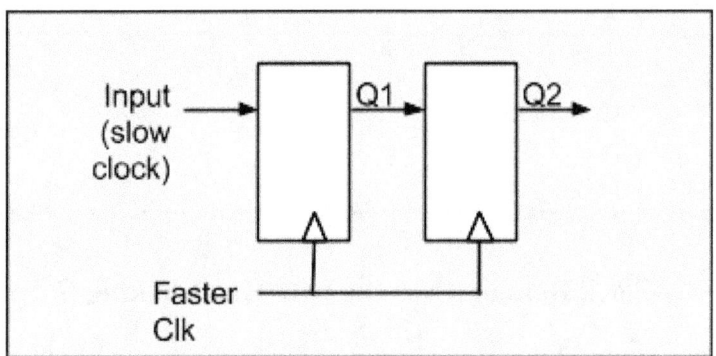

**Classic 2-Stage Synchronizer** *(Fig. 21)*

Why are 2 stages needed? The first Q1 flop can become metastable since its input is driven from a different clock. A second stage allows the Q1 output to settle. The Q2 output will be stable and can be used. STA tools will not check the input path to Q1, but it is must check the Q1 to Q2 path.

## 33. Design a circuit to synchronize a signal from a fast domain to a slow domain?

The circuit still requires a 2-stage synchronizer between the fast domain (source) and the slow domain (destination). However, the faster source domain needs extra control logic to make sure the signal is held stable until the slow domain has a chance to synchronize it. This requires a handshaking signal from the slow domain back to the fast domain (through yet another synchronizer). Look at the schematic below:

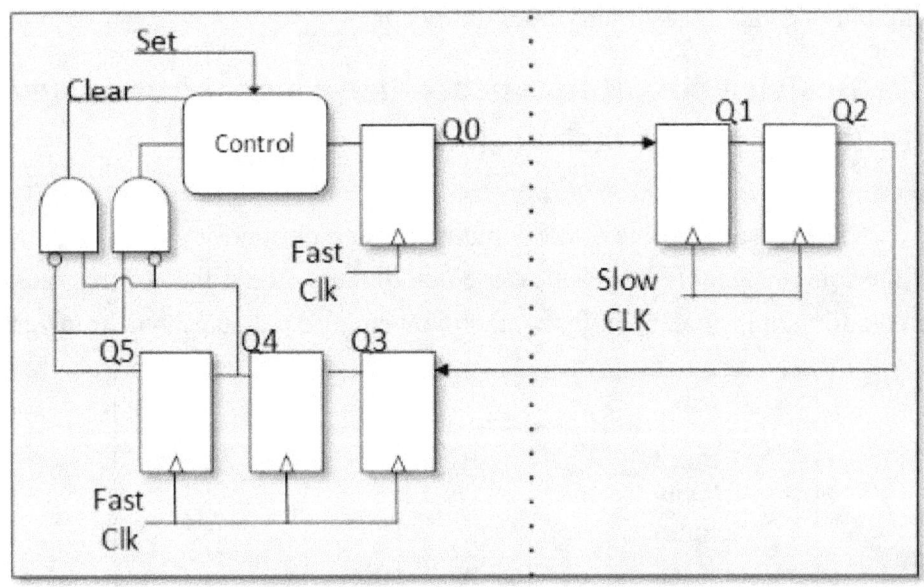

**Clock Domain Crossing with Handshaking** *(Fig. 22)*

Once Q0 has been set, the control logic is responsible for holding the Q0 output steady (not allowing it to change) until it sees the handshake response from the other domain (in this case a rising edge pulse between Q4 and Q5), which indicates that the Q2 flop has captured the signal. Only then can the Q0 flop be cleared, but the control logic should prevent the setting of Q0 again until it sees a falling edge between Q4 and Q5 (which indicates that Q2 is low again) and the circuit can safely start over.

# 34. How would you synchronize a data bus (multi-bit)?

If you need to synchronize a multi-bit bus, you should not use a classic double flop synchronization approach as previously mentioned (unless the bus has first been Gray encoded) . The problem is some of the bits will make it across in 1 cycle and some in 2 cycles (due to the timing, and process variations of the silicon). Therefore a guaranteed approach is to only synchronize a 1-bit control signal to indicate to the other domain that the bus is ready to be synchronized (that is, the bus is stable and will not change). The control signal should be synchronized with double flop as before and then used as a mux select to allow the new data to come across (otherwise the old value is held). Below is diagram of the circuit:

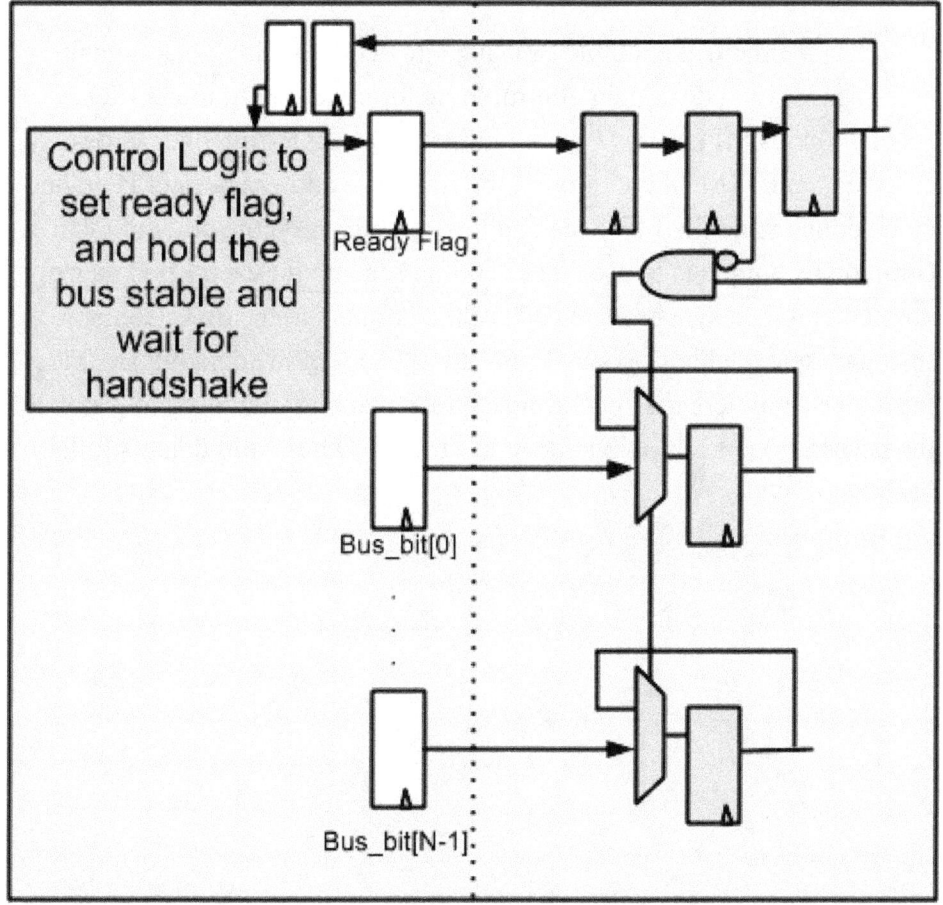

**Synchronize Bus Across Clock Domains (Full Handshake)** *(Fig. 23)*

# 35. Why are Gray coded data buses ok to synchronize using classic double flop approach on all the bits?

Gray coding by definition is an encoding scheme where only 1 bit changes between states. Since only 1 bit is changing, then it is ok to synchronize a bus from one domain to another domain using just the classic double flop synchronizer scheme. As you remember the problem with classic synchronization (double flop) across a bus is that some of the bits may transition before others (some flops would meet setup/hold condition) while others may not. But in the case where only 1 bit is changing, you don't care about this issue since only 1 bit is changing, and therefore the correct value eventually makes it across safely.

One practical example of using Gray coding with clock domain crossings is when designing control logic for an asynchronous FIFO. Asynchronous FIFOs are very common when transferring data across clock domains. Most commonly, the fifo write address pointer is on one clock domain and the FIFO read address pointer is on the other domain. A common technique is to use a gray coding scheme for each FIFO address pointer, and then synchronize each of the pointers to the other domain (using classic double flop synchronization on all of the data bits). Once the address pointers are synchronized, they can be safely used to generate the necessary control logic to decide if the FIFO has valid data, is full, or is empty.

# Power Related Questions

# 36. Describe the two components that make up power

  The two components of power in digital circuits are *static* and *dynamic* power. Low Power RTL design techniques are very important in today's digital logic designs because battery longevity plays such an important role in mobile consumer electronics and also with wearables.

# 37. What is static power and how to reduce it in RTL?

  Static power, also known as leakage power, is directly related to the size or area of your design. The area is directly related to RTL code which creates logical gates (and are made up from transistors). When power is applied to the transistor, the transistors naturally leak current due to the manufacturing defects and physical characteristics of the silicon.. In order to reduce static power, you can either minimize the number of gates in your design by using resource sharing techniques in the RTL code, or by literally turning off the power to that logic.

  A power controller is used to request power or turn off power depending upon if that logic will be used. Usually large sections of transistors are grouped together in their own power domain when power is removed when that logic is not being used. When sections of logic are powered down it is important that *isolation cells* are placed on all the outputs of the powered off circuits to ensure that a known value is driven after the power is turned off. Isolation cells are special cells that could be instantiated in the RTL code (or inserted in the design by the back end physical design team). It is important for the designer to make sure the correct isolation value is driven (either high or low) to ensure the circuit stays in an inactive state (usually driven to same as reset value of the flop).

  When power is turned off to the digital logic gates and registers, the values are lost. However, it may be desired for certain important control registers that you want to keep (or retain) its value when power is turned off. There types of flops are called *retention flip flops*.

# 38. What is dynamic power?

Dynamic power is the power that is lost when a signal transitions from a logic high-to-low state or from a logic low-to-high state. This is because today's silicon circuits are made using CMOS transistors. When a signal transitions, there is a current spike that occurs when the transistors are switching from an on to off state (for a very fast moment there is a path from VDD to GND). The dynamic power lost is a direct result of the amount of transitioning and toggling of signals inside your code.

# 39. Describe some low power RTL coding techniques

Dynamic power is easy to reduce in your RTL code by avoiding the unnecessary toggling of signals. This can be accomplished by using an enable signal to qualify when D <= Q. Any datapaths in the code should use enable signals if possible. Another example of using an enable signal is with counters. The below code ensures the counter doesn't toggle unless it's time to start and doesn't toggle after it is time to stop:

```
always @(posedge clk or negedge rst)
if (!rst)
  counter <= #1 {8{1'b0}};
else if (start_counter == 1'b1)
  counter <= #1 {8{1'b0}};
else if (counter < 8'b00101111)
  counter <= #1 counter + 1;
```

**Start/Stop Conditions for Counters** *(Ex. 28)*

By using an enable signal, registers will not toggle unnecessarily. Also, synthesis tools understand enable coding styles and automatically insert clock gating cells for you automatically (using the enable signal to turn on and off the clock). These are considered leaf level clock gates.

The RTL design could also instantiate its own clock gating cells at a middle level (above the leaf level). This can be useful when the designer knows that a group of logic will not be used (for example a feature enabled or disabled with a software programmable register bit). If you design your own clock gating cell, be sure it doesn't glitch (refer to interview question about how to code a clock gating circuit). You should not add a clock gate cell for a small number of flops (because adding more logic will increase the static power. You should consider this tradeoff).

When trying to reduce dynamic power, one method is to only toggle 1 bit of your FSMs as it transitions between states. This can be accomplished with Gray coding for the state transitions. Gray coding can also be used in any read or write address busses used with internal FIFOs or memories.

Another method to reduce dynamic power is to use databus inversion techniques. In some applications, if you have a wide databus you may want to reduce the amount of individual bit toggling. Instead of driving each bit to its correct high or low value, in some cases it makes sense to keep the bus in its previous logic state and assert a sideband signal called *inversion* which indicates to the receiving logic that the databus bits are inverted. This approach is valuable if the number of bits which are different between the current value and the next value of the databus are more than N/2 (where N is the width of the bus).

There are plenty of white papers on this subject if you search on low power verilog databus encoding schemes on Google.

**Author's Tip**: This was a high level introductory discussion on dynamic and static power and RTL reduction coding guidelines. The subject matter is much deeper and if you are specifically interviewing for a low power design position you should dive much deeper into the material.

# Digital Logic Design
# Interview Questions

# 40. What is definition of setup and hold time?

The setup time for a flip flop is the minimal amount of time its input signal must be arrive and be stable (not toggle) before its clock edge. Hold time is the minimal time after a clock rises that data must remain stable and not change (hold steady).

One common reason setup times are violated is because there is too much logic between the launch flop and capture flop for the desired clock frequency to be met. The capture flip flop would have a setup time violation if the delay from the launch flop (clk->q + wire delay + propagation delay through combo logic) was greater than the clock period.

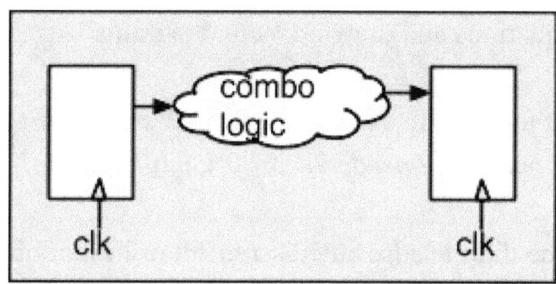

**Example of Sequential Circuit** *(Fig. 24)*

There are multiple ways to fix setup violations. If it is early enough in the design cycle you could rewrite the RTL code (micro architect it in a different way). Or you could add a pipeline stage in the RTL path (break the path in the middle somewhere by adding another flop stage) and use an enable signal at the destination capture flop. Additionally the physical design team could resize the cells to have greater drive strength, or move the physical location of the cells closer to each other. And as a last resort, you could always run the clocks slower.

A hold violation is to the opposite extreme of setup violations. If very little logic existed between the launch and capture flops, the signal may arrive too soon and cause a hold violation. To fix a hold violation, the physical design team could add delay in the path (for example buffers).

# 41. Venn Diagrams and Boolean Logic

When I first started interviewing, I was hoping to rely on my practical work experience to get me through the job interview questions. I was quickly proven wrong. In fact, I was surprised on how college like the questions were. For example, I have only written RTL code (and never once had to draw a Karnaugh map). However, as I learned the hard way, you should be prepared to answer Karnaugh maps questions on interviews!

On one interview I was introduced to a senior design engineer. He asked me if I was a digital designer, and I nodded my head yes. He quickly followed up "then you shouldn't have any problem solving boolean logic equations and drawing Venn diagrams".

I thought to myself, *"Huh? I've been designing digital circuits for years and never once did I need to draw a Venn Diagram!"*

Hoping he didn't sense any discomfort or hesitation on my part, I quickly said that I haven't done this since my digital logic classes in college. He didn't care, and proceeded to draw on the whiteboard. He turned to me and asked:

"What is the boolean equation for this shaded region below?"

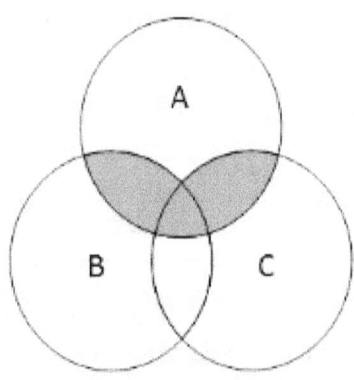

**Venn Diagram** *(Fig. 25)*

I studied the diagram for couple of seconds. I could see it was made up of three intersections: AB || ABC || AC. I noticed that the middle cross section of ABC was redundant and was already covered by the other terms, so without thinking any longer I blurted out "AB *or'd* with AC", and I wrote on the board: AB || AC.

I pointed out that the middle cross section of (ABC) was redundant and already covered in my equation.

He nodded, and then asked me if I could simplify my equation even further. I looked at my equation again, and quickly answered, "yes, you can factor out the A", and then I wrote:

$$AB \parallel AC = A (B \parallel C)$$

He then asked if I could draw the equivalent logic gates so I drew:

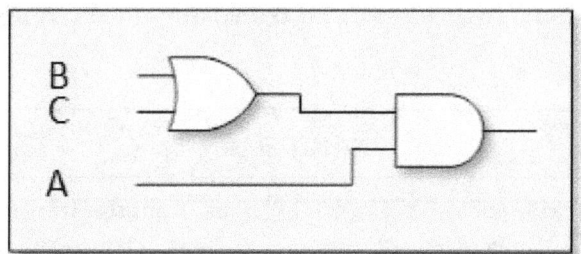

**Logic Gates for A&(B|C)** *(Fig. 26)*

Again, he gave a quick nod of the head for agreement. Then he asked if I could draw the transistor cell equivalent. I turned my head slowly and looked at him as if to ask "are you for real?" He responded that he just wanted to see how far I could go. I responded that once upon a time I could draw a CMOS inverter, and given some time I could probably figure out the other gates.

---

**Author's Tip:** In reality, I should have been prepared for this question if I really wanted to impress him. On the next few pages, I've included the CMOS transistor level equivalent for the common digital logic gates. You should memorize all of these circuits and be prepared.

---

# 42. Transistor Level Equivalent of Digital Logic Gates

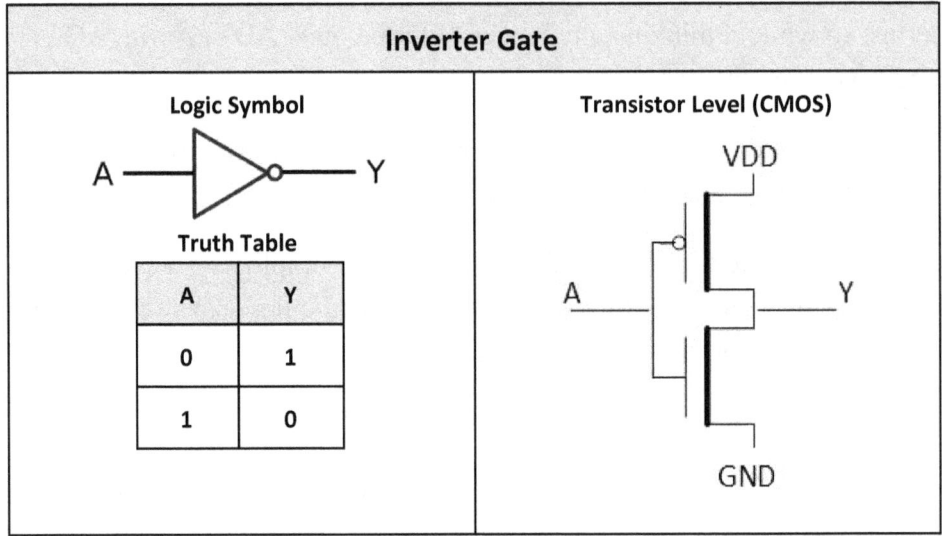

Inverter Gate, Truth Table, and transistor-level circuit *(Table 2)*

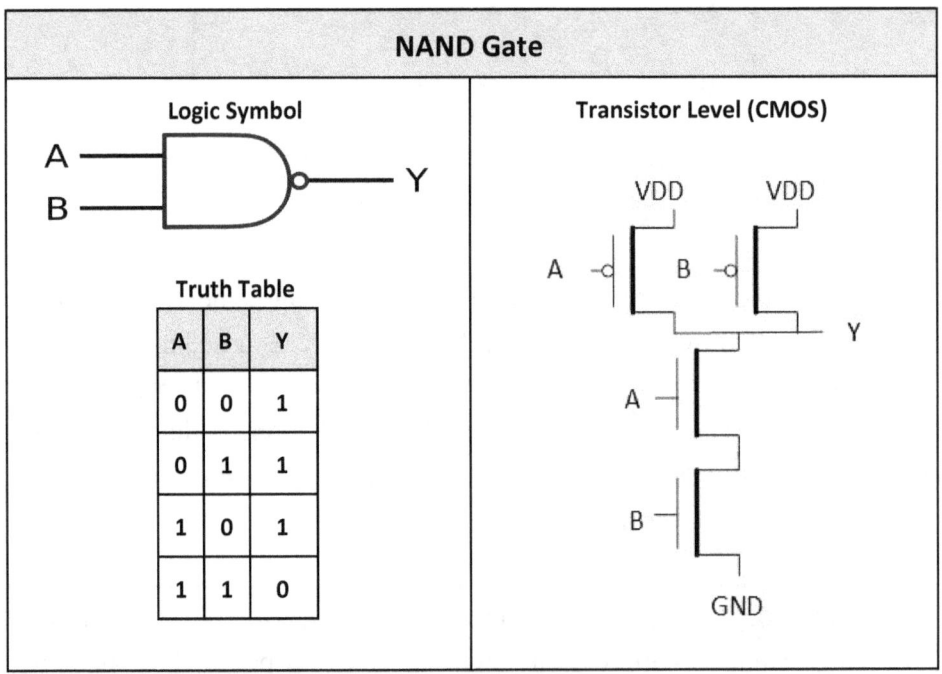

NAND Gate, Truth Table, and transistor-level circuit *(Table 3)*

**NOR Gate**

Logic Symbol

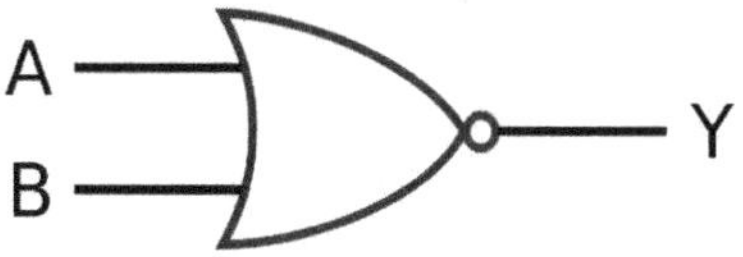

Truth Table

| A | B | Y |
|---|---|---|
| 0 | 0 | 1 |
| 0 | 1 | 0 |
| 1 | 0 | 0 |
| 1 | 1 | 0 |

Transistor Level (CMOS)

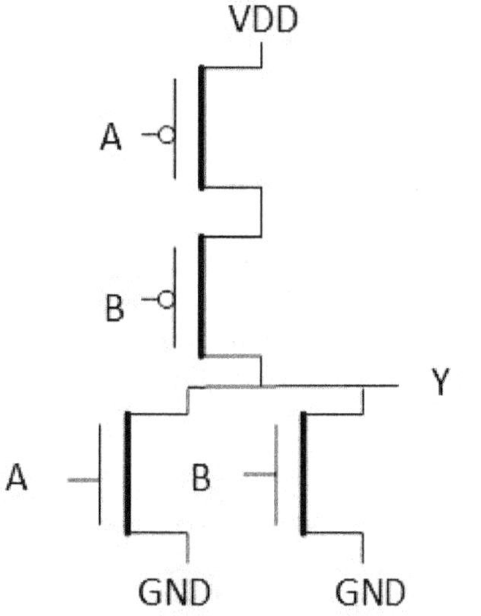

**NOR Gate, Truth Table, and transistor-level circuit** *(Table 4)*

**AND Gate (created from a Nand and Inverter)**

Logic Symbol

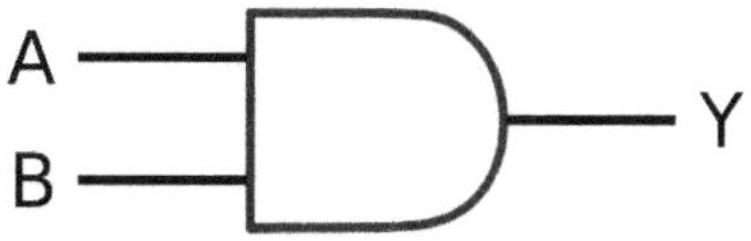

Truth Table

| A | B | Y |
|---|---|---|
| 0 | 0 | 0 |
| 0 | 1 | 0 |
| 1 | 0 | 0 |
| 1 | 1 | 1 |

Transistor Level (CMOS)

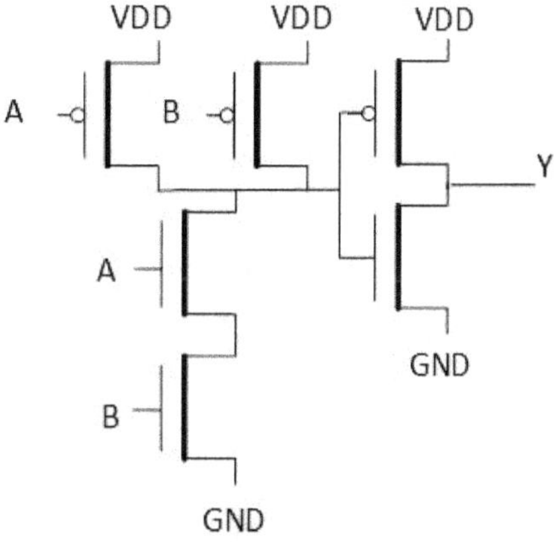

**AND Gate, Truth Table, and transistor-level circuit** *(Table 5)*

# OR Gate (created from a Nor and Inverter)

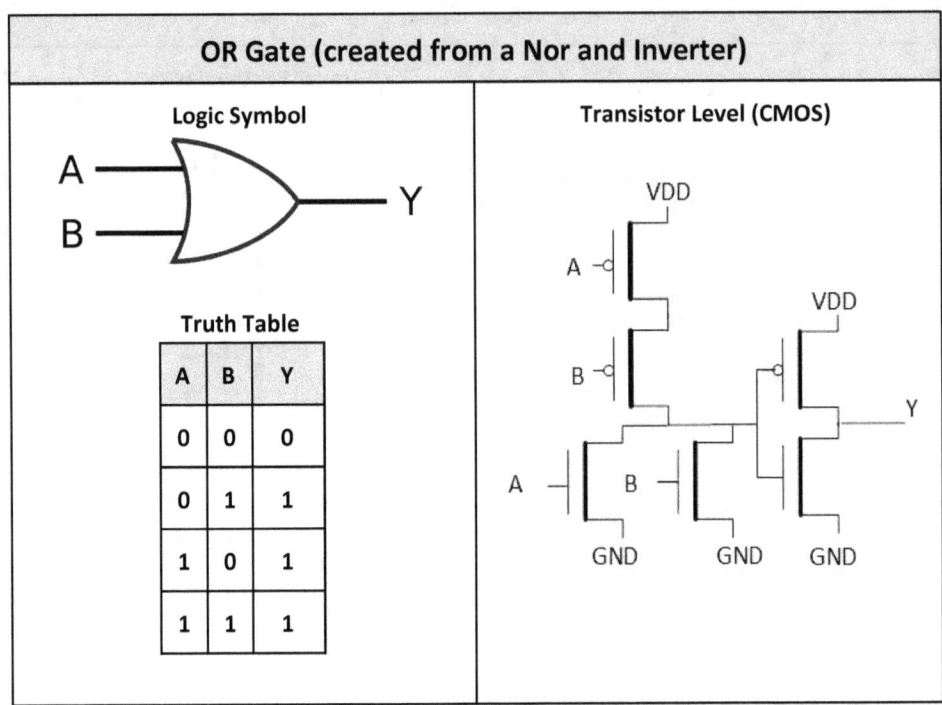

| Logic Symbol | Transistor Level (CMOS) |

**Truth Table**

| A | B | Y |
|---|---|---|
| 0 | 0 | 0 |
| 0 | 1 | 1 |
| 1 | 0 | 1 |
| 1 | 1 | 1 |

**OR Gate, Truth Table, and transistor-level circuit** *(Table 6)*

# 43. Draw the cross section of a CMOS transistor

When I first started interviewing, I was not prepared for the many types of academic questions. I had been out of school for many years and quite frankly forget the information that I didn't use daily. Despite always introducing myself as a front end RTL design engineer, I was still expected to know the fundamental engineering concepts.

On one interview, I was asked to draw a cross-section of a CMOS transistor. I *vaguely* remember some details from my Semiconductor course in college, but I restated that my primarily focus was writing Verilog code. This was obviously *not* the right answer! For your reference, below is the cross-section of a CMOS transistor. I suggest simply memorizing it and be ready to draw it and explain it:

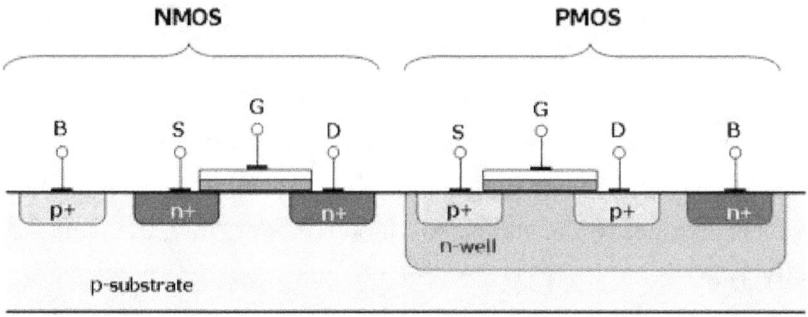

**Cross-section of CMOS Transistor** *(Fig. 27) Image from Wikipedia.org*

You should also be familiar with process nodes and know in which technology nodes your designs were in. Refer to table below:

| Year and Process | | Year and Process | | Year and Process | |
|---|---|---|---|---|---|
| 1989 | 800 nm | 2001 | 130 nm | 2012 | 22 nm |
| 1994 | 600 nm | 2004 | 90 nm | 2014 | 14 nm |
| 1995 | 350 nm | 2006 | 65 nm | 2016 | 10 nm |
| 1997 | 250 nm | 2008 | 45 nm | 2018 | 7 nm |
| 1999 | 180 nm | 2010 | 32 nm | 2020 | 5 nm |

**Semiconductor Manufacturing Processing Nodes** *(Table 7)*

## 44. FSMs, Karnaugh Maps and Gray Code

Not once in my professional career did I need to draw a Karnaugh map. So I was a bit surprised when one interviewer directly asked me to do so. I was not simply not prepared.

On this particular interview, I knew beforehand (from the recruiter) that I should understand how clock dividing circuits worked. I prepared myself on how to design any clock divide by N circuit (with a 50% duty cycle).

The interviewer asked me to "design a clock divide by 3 circuit using a FSM (finite state machine)". He also stated that there was not a requirement for duty cycle to be 50% (which simplifies the question).

I said I would create a FSM with 3 states, and in state 1 I would drive high output, and the other two states I would drive a low clock output. As I drew the state diagram on the whiteboard, he asked me to assign binary values to the states, so I labeled them "00", 01", and "10" respectively:

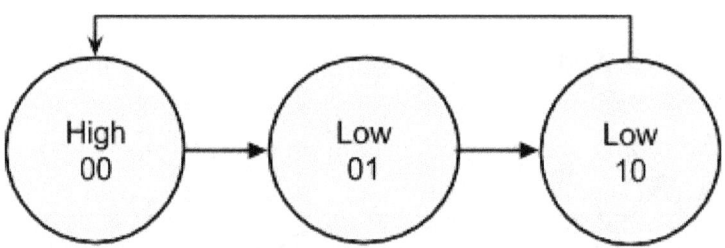

**3 State FSM for Clock Divider** *(Fig. 28)*

He nodded in agreement. He then asked me to change the state values that I assigned to instead use gray coding. I understood that Gray coding rule (only 1 bit could change between each state transitions). So I relabeled the states just as he asked, with the following values: 00, 01, 11. As soon as I assigned the last state, I realized there would be a problem going from the last state back to the first state since two bits changed, violating Gray code rules.

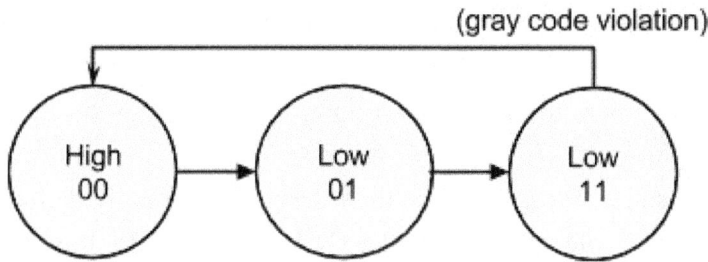

**Gray coding (with violation from state 11 to state 00)** *(Fig. 29)*

This is when I got stuck (and then nervousness set in). After some time, he asked me to draw a Karnaugh map for the state values. At this point of the interview, I really felt I was following a rehearsed script, like he had asked this question hundreds of times before and he wanted to see how my thought process would work. As we worked through this problem together, I draw the following Karnaugh map for my FSM:

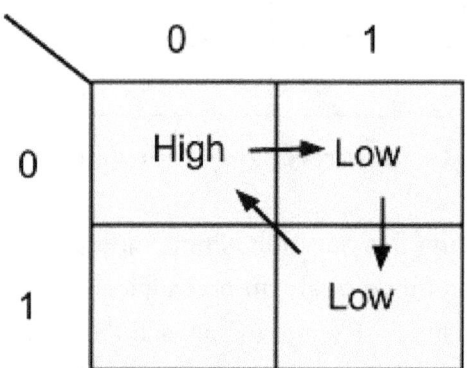

**2-bit FSM (illegal gray code transition)** *(Fig. 30)*

Clearly from the low state 11 it is illegal to go back to high state 00. In the figure above, there is an empty slot (state 10) available which could be used to generate a high output. If you were to use this unused state, then the FSM could reverse and go back to two previous low states.

That would require extra control logic (flags) to be set, but that solution did not occur to me at the time..

He reminded me that Karnaugh maps squares are structurally already gray coded. That is, each adjacent square in a Karnaugh map (above or below, or left or right) is always a valid gray code transition. This is a very important to remember and I had forgot this since college.

After some discussion with the interviewer, I finally deduced that I could use a 3-bit state value for the FSM. He asked me to draw a new Karnaugh map to represent the new states. I decided to use the following states and start in the upper left corner: 000 (high), 001 (low), 011 (low), then 111 (high), 101 (low), and 100 (low):

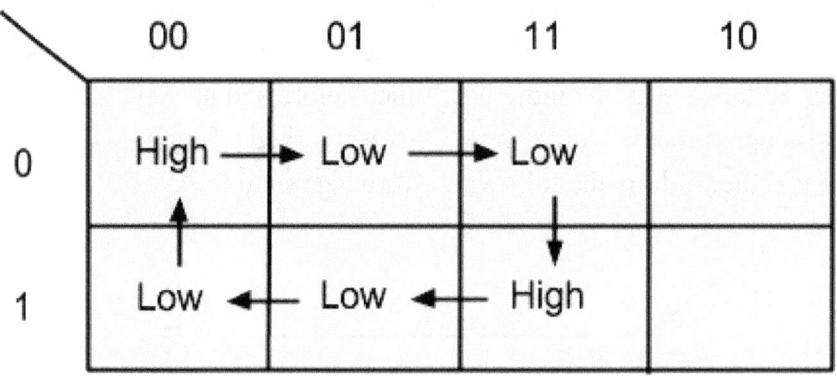

**Six state FSM with correct gray code values** *(Fig. 31)*

The key to solving this question is remembering that when using Karnaugh maps you can move freely around adjacent state spaces with only a 1-bit change in value. By moving around the states, you can piece together the answer. In fact, any six states will work as long as you circle back to the beginning. After this interview, Gray coding started to become a recurring theme in many of my interviews questions.

Gray coding techniques are discussed in other questions in this book with other possible use-case scenarios (CDC and low power).

# 45. Half Adder using XOR gates and AND and OR gates

A basic question and concept that you should understand is how to design digital logic for adders. Below is the truth table, K-map, and logic implementation for an adder:

| A | B | Result | Carry |
|---|---|--------|-------|
| 0 | 0 | 0 | 0 |
| 0 | 1 | 1 | 0 |
| 1 | 0 | 1 | 0 |
| 1 | 1 | 0 | 1 |

**Truth Table for an Adder** *(Table 8)*

Karnaugh Map for the Result and Carry bits:

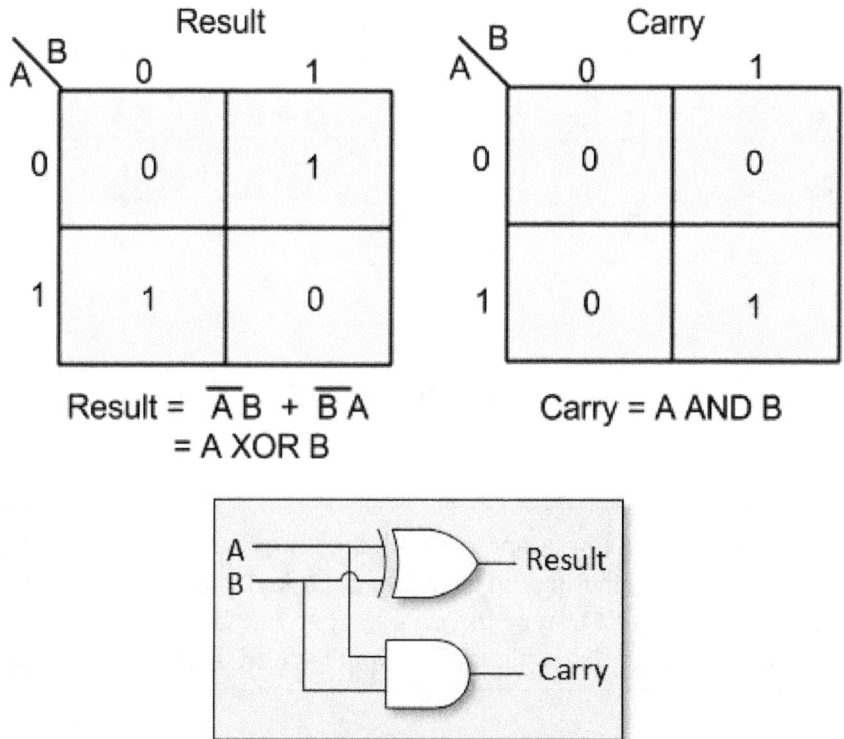

Result = $\overline{A}B + \overline{B}A$
= A XOR B

Carry = A AND B

**Half Adder** *(Fig. 32)*

By placing two half-adders together you can create a full adder circuit:

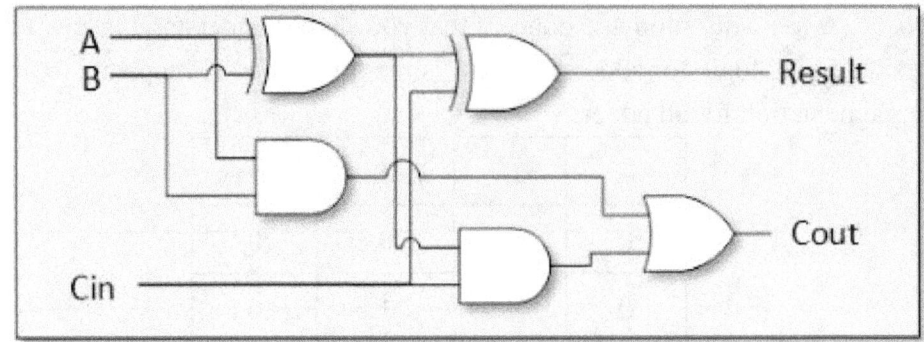

**Full Adder Circuit** *(Fig. 33)*

Placing multiple 1-bit full adders together and you can get a multibit adder. Below is a 4-bit adder:

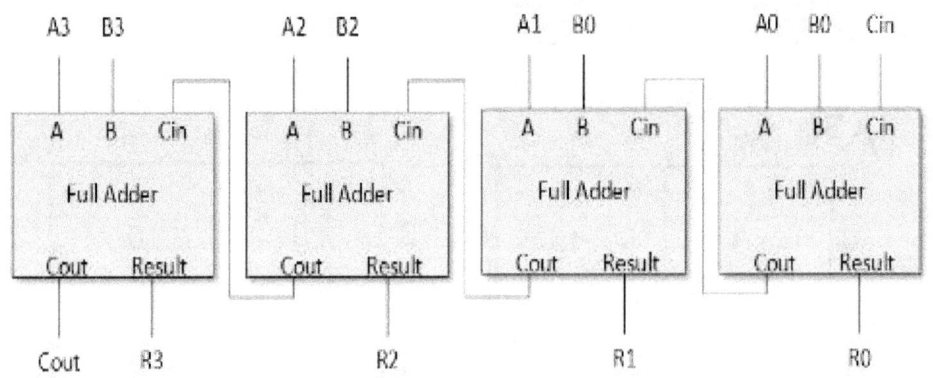

**Multibit Adder** *(Fig. 34)*

**Author's Tip:** It is important to understand how these basic building blocks are used together just in case you are asked. It is too easy as a front end designer just to type in verilog: R = A + B; Now you will have understanding what's happening in the synthesis tool.

# 46. Creating Digital Logic Gates using 2:1 Mux

You'll be surprised to know the following questions appeared more often than not. I consider these questions tricks, and they are not practical at all. However, it does make you understand the truth table of the common gates and also are logical once you undersatnd you can use VDD and GND when you construct the circuit. You need to create all the logic gates using using a 2-input mux. I suggest you just memorize this.

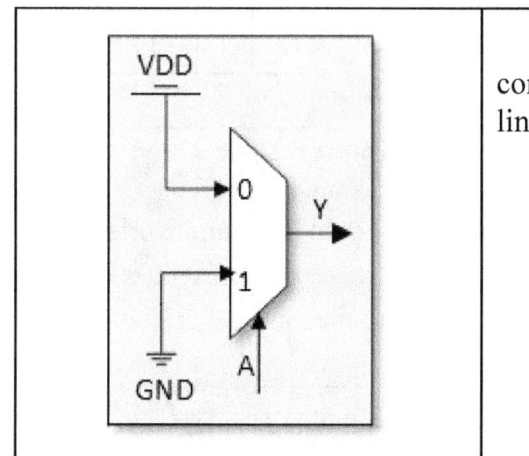

For an *inverter*, you should connect the A input to the select line (A=S)

| A | Y |
|---|---|
| 0 | 1 |
| 1 | 0 |

**Inverter Implemented with 2:1 mux** *(Table 9)*

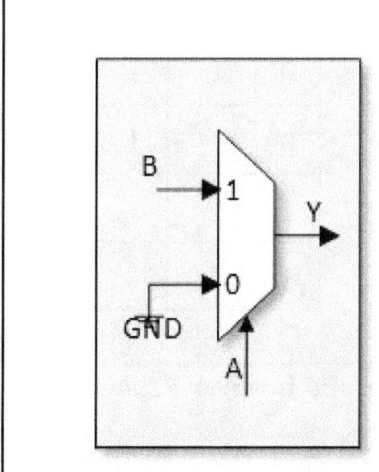

For an *AND* gate, you connect the A input to the select line (A=S)

| A | B | Y |
|---|---|---|
| 0 | 0 | 0 |
| 0 | 1 | 0 |
| 1 | 0 | 0 |
| 1 | 1 | 1 |

**AND Gate Implemented with 2:1 mux** *(Table 10)*

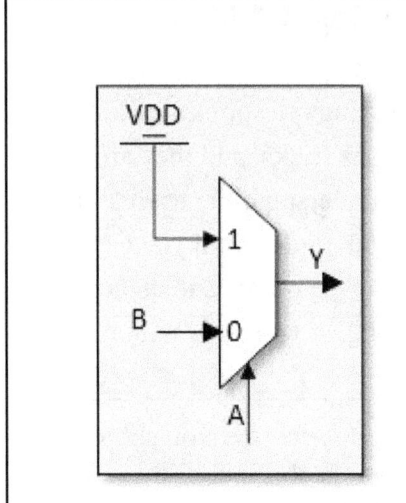

| For an *OR* gate, you connect the A input to the select line (A=S) | | |
|---|---|---|
| **A (S)** | **B** | **Y** |
| 0 | 0 | 0 |
| 0 | 1 | 1 |
| 1 | 0 | 1 |
| 1 | 1 | 1 |

**OR Gate Implemented with 2:1 mux** *(Table 11)*

To create a NAND or NOR gate, you could place an inverter on the output of and AND or OR gate. Another method is presented below in which you could invert the A or B input and then use the 2:1 mux implementation below:

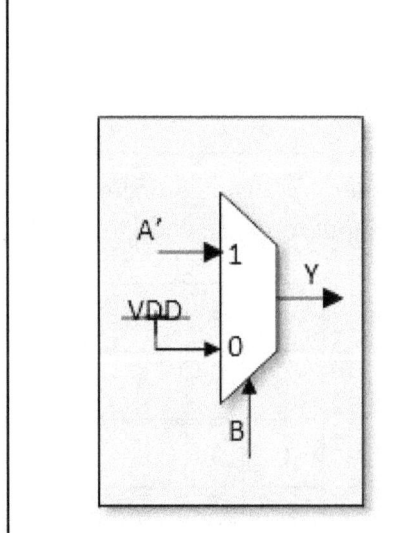

| For an *NAND* gate, you connect the B input to the select line (B=S), and then use A' (not A) | | |
|---|---|---|
| **A** | **B (S)** | **Y** |
| 0 | 0 | 1 |
| 0 | 1 | 1 |
| 1 | 0 | 1 |
| 1 | 1 | 0 |

**NAND Gate Implemented with 2:1 mux and inverter** *(Table 12)*

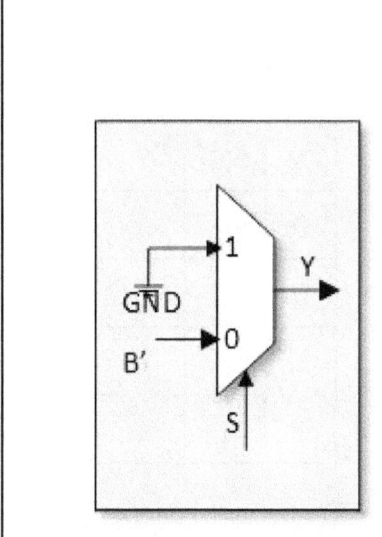

For an *NOR* gate, you connect the A input to the select line (A=S), and then use B' (not B)

| A (S) | B | Y |
|---|---|---|
| 0 | 0 | 1 |
| 0 | 1 | 0 |
| 1 | 0 | 0 |
| 1 | 1 | 0 |

**NOR Gate Implemented with 2:1 mux and inverter** *(Table 13)*

Along with an inverter, you could also create an XOR gate with 2:1 mux:

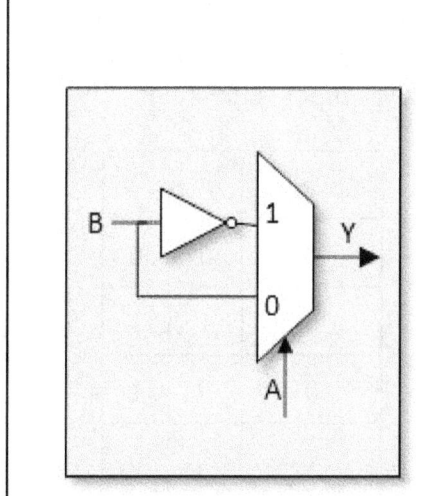

For an *XOR* gate, you connect the A input to the select line (A=S)

| A (S) | B | Y |
|---|---|---|
| 0 | 0 | 0 |
| 0 | 1 | 1 |
| 1 | 0 | 1 |
| 1 | 1 | 0 |

**XOR Gate Implemented with 2:1 mux and inverter** *(Table 14)*

And for completeness, the XNOR gate with 2:1 mux and inverter is included here:

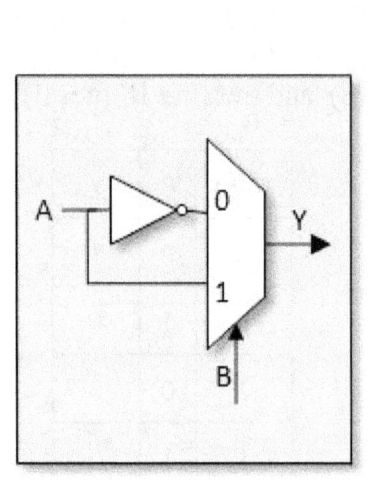

For an *XNOR* gate, you connect the B input to the select line (B=S)

| A | B(S) | Y |
|---|------|---|
| 0 | 0 | 1 |
| 0 | 1 | 0 |
| 1 | 0 | 0 |
| 1 | 1 | 1 |

**XNOR Gate Implemented with 2:1 mux and inverter** *(Table 15)*

## 47. How do you use and XOR gate as a controlled inverter?

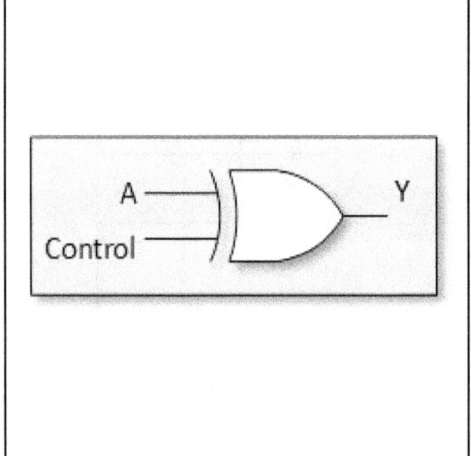

When *control* is 0, the A input passes through. When control is 1, the A input is inverted.

| Control | A | Y |
|---------|---|---|
| 0 | 0 | 0 |
| 0 | 1 | 1 |
| 0 | 0 | 1 |
| 1 | 1 | 0 |

**XOR Gate Used as Controlled Inverter** *(Table 16)*

# 48. Design an *INVERTER, AND, OR*, and *XOR* Gate using only *NAND* Gates

I'm not sure why these questions are asked, since they are not practical. But, here are some common solutions to creating other gates with just nand gates.

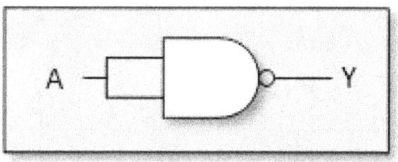

**Inverter** *(Fig. 35)*

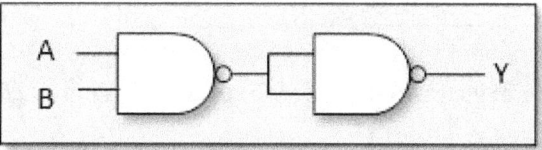

**AND Gate Implemented with NAND Gates** *(Fig. 36)*

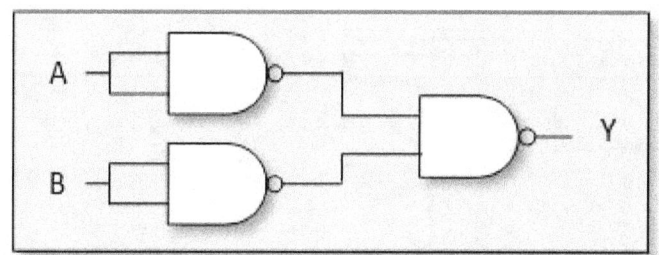

**OR Gate Implemented with NAND Gates** *(Fig. 37)*

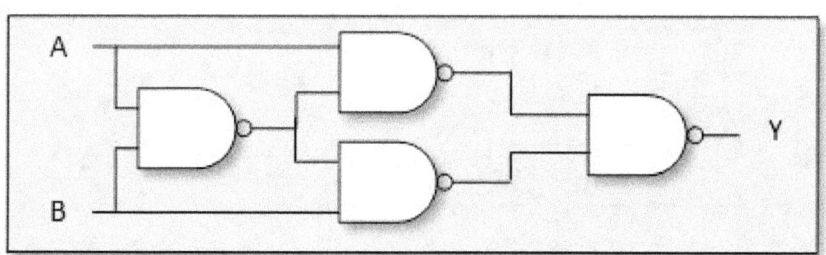

**XOR Gate Implemented with NAND Gates** *(Fig. 38)*

## 49. Create a 4:1 Mux using 2:1 Muxes and an Inverter

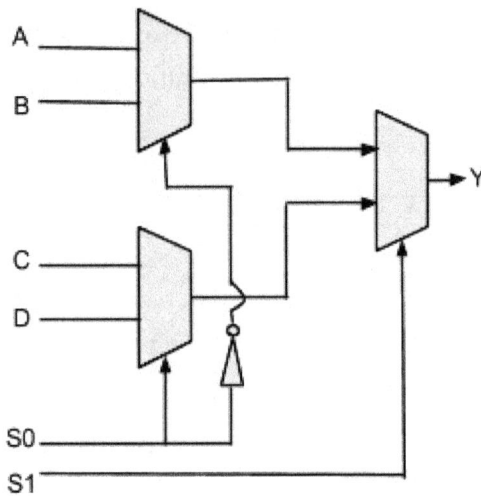

**4:1 Mux created with 2:1 muxes and inverter** *(Fig. 39)*

Can you design a 4:1 mux with only combinatorial logic?

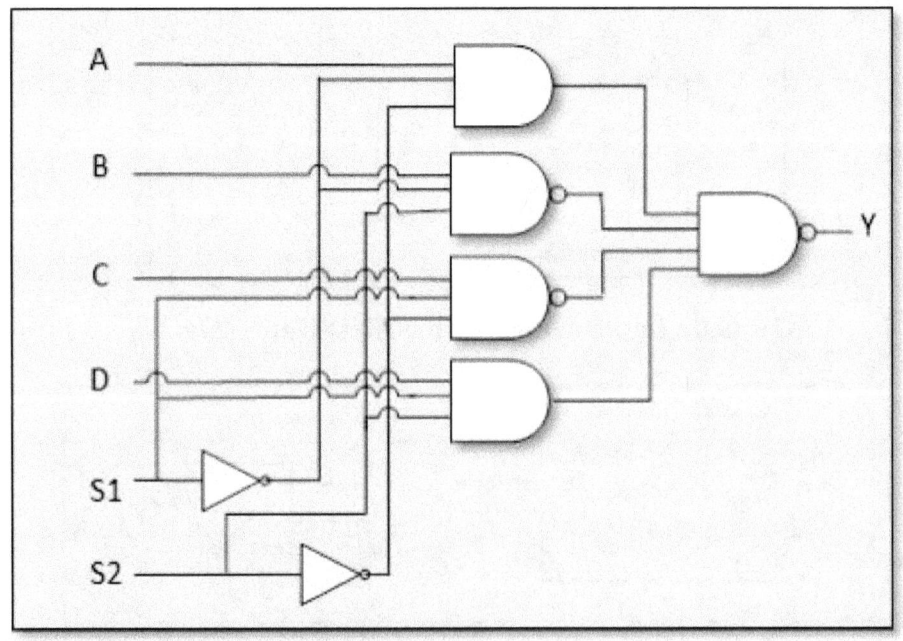

**Example 4:1 mux implemented with combinational logic** *(Fig. 40)*

# 50. Frequency, Period, and Propagation Delay

Assuming propagation delay through each gate below is 1 *ns*, what is the maximum frequency this circuit can run?

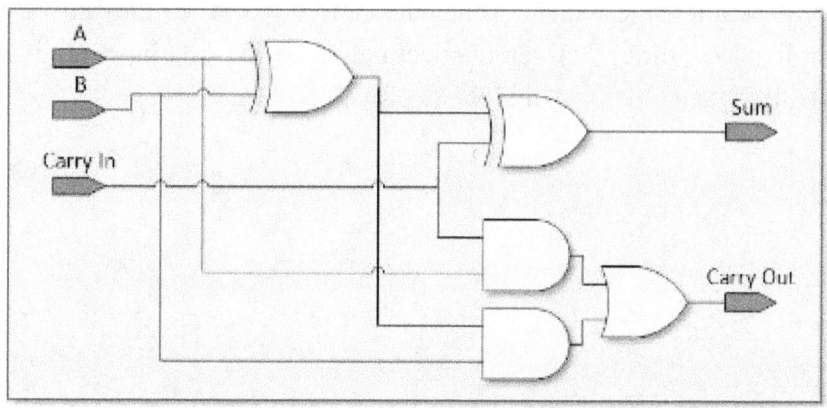

***What is maximum frequency this circuit can operate?*** *(Fig. 41)*

Frequency = 1 / T. The maximum frequency is directly related to worst total path in the circuit. In this example, the worst path is a 3 gate delays starting from point A through the *XOR* gate, then *AND* gate, then the *OR* gate for a total path delay of 3 *ns*. Therefore, the safest (fastest) frequency is 333.33 megahertz ( 1/ 3 *ns*). As a reminder, there are 4 possible timing path groups in a given module: input to output port, input ports to internal flop, internal flop to internal flop, and internal flop to output port.

---

**Author's Tip:** Remember that in a real silicon circuit, there are also wire delays associated with the signals. In addition, each of the inputs will have some an input delay associated with them (even if they are driven from flops there is a clock to Q delay before the signal becomes available and routing). For output ports, keep in mind there is a capture flop on the other end so you must meet the setup time requirement of the capture flop.

---

# 51. Convert the Decimal Value 13 to Binary, Hex, and Octal Values

There is nothing tricky about this problem. It is straightforward but some people have issues for the conversion to octal (if they haven't done it in a long time). Given the decimal value 13, the binary equivalent is 1101, the hex value is D, and the octal value is 15.

# Logical Problem Questions

When I'm traveling for a job interview and have extra time to pass, I like to entertain myself by working logical puzzles. Logic puzzles can help stimulate your brain and may prepare you for interview questions.

They may not seem related to digital design at first, but they do require logical thinking to solve them (which is the basis of engineering). Whether these types of problems detect if a candidate would be a good employee or not is arguable. I think it is better to be prepared in case these types of questions are asked. While I have not seen any, some of my colleagues have been asked these types of questions or variations of them.

The questions listed in this section were found on the internet just by searching for logic puzzles. I selected a few that do not have trick answers or tricky wording. I've only included questions that have logical solutions. Also, I did not include any intensive math related questions or any that require advanced mathematical calculations. Instead, these questions involve straightforward logical thinking and problem solving skills.

## 52. Four Gallons of Water

You have 2 empty buckets: a 5 gallon bucket and a 3 gallon bucket. How can you measure exactly four gallons of water and put it in the 5 gallon bucket? (Assume that you have an unlimited supply of water and that there are no measurement markings of any kind on the buckets. You can't estimate either, for example, fill it halfway is not acceptable.)

## 53. The Path to Freedom

You are a prisoner in a cold dungeon. After a long search you discover three doors. Behind only one of the doors is the path to freedom. Behind the other two doors, however, is not the path to freedom (you can assume some violent death ensues…). Each door has an inscription above them. They are labeled as follows:

| Door #1 | Door #2 | Door #3 |
|---------|---------|---------|
| This door leads to freedom | This door does not lead to freedom | The second door does not lead to freedom |

**Path to Freedom Doors** *(Table 17)*

Given the fact that at *least one* of the three statements on each door is true and at *least one* of them false, which door would lead to freedom?

## 54. The Three Light Switches

There are three light switches in your basement. All of them are currently in the off position. Each switch controls its own lamp on the floor above. You need to find out which light switch controls which lamp. You are not allowed to break walls or inspect wiring of circuits, etc.

You must start in the basement where the three light switches are located. You may turn on or off the switches any number of times, but you may only go upstairs one time to inspect the lamps (and at that time you must make your decision). How can you determine which switch controls each lamp with only one trip upstairs? *Hint: you may need to think outside the box for this one, but the answer is still logical.*

## 55. Multiply Question

|  |  |
|---|---|
| ABCD<br><br>x    9<br>———<br>DCBA | What are the values for A,B,C,and D?<br><br>*(Eeach variable represents a different single digit value between 0 and 9)* |

**Multiply Question** *(Table 18)*

# 56. Einstein's Riddle

This is a classic logical question which is attributed to Einstein. . Regardless of its disputed origin, many people are not able to answer it.

There are 5 houses in 5 different colors in a row. In each house lives a person with a different nationality. The 5 owners drink a certain type of beverage, smoke a certain brand of cigar, and keep a certain pet. No owners have the same pet, smoke the same brand of cigar, or drink the same beverage. Other facts:

1. The Brit lives in the red house.
2. The Swede keeps dogs as pets.
3. The Dane drinks tea.
4. The green house is on the immediate left of the white house.
5. The green house's owner drinks coffee.
6. The owner who smokes Pall Mall rears birds.
7. The owner of the yellow house smokes Dunhill.
8. The owner living in the center house drinks milk.
9. The Norwegian lives in the first house.
10. The owner who smokes Blends lives next to the one who keeps cats.
11. The owner who keeps the horse lives next to the one who smokes Dunhill.
12. The owner who smokes Bluemasters drinks beer.
13. The German smokes Prince.
14. The Norwegian lives next to the blue house.
15. The owner who smokes Blends lives next to the one who drinks water.

**Question:** Who owns the fish?

# *LOGICAL PROBLEMS (Answers)*

# 52. Four Gallons of Water
**ANSWER**:

1. Fill up the 5-gallon bucket.
2. Pour the contents of this into the 3-gallon bucket. You are now left with two gallons of water in the 5-gallon bucket.
3. Dump out the water in the 3-gallon bucket.
4. Pour the two gallons of water that are in the 5-gallon bucket into the 3-gallon bucket.
5. Fill up the 5-gallon bucket up again.
6. Top off the 3-gallon bucket with water from the 5-gallon bucket leaving you with 4 gallons of water in the 5-gallon bucket.

**ALTERNATE ANSWER**:

1. Fill the 3-gallon bucket.
2. Pour the 3 gallons of water into the 5-gallon bucket
3. Fill the 3-gallon bucket again.
4. Fill up the 5-gallon bucket with the 3-gallon bucket, leaving you with 1 gallon left in the 3-gallon bucket.
5. Empty out the 5-gallon bucket.
6. Pour the remaining 1 gallon of water from the 3-gallon bucket into the 5-gallon bucket.
7. Fill the 3-gallon bucket.
8. Pour the 3 gallons of water from the 3-gallon bucket into the 5-gallon bucket leaving you with 4 gallons of water in the 5-gallon bucket.

# 53. The Path to Freedom:
**ANSWER**: Door #3

1. If you assume Freedom is behind the first door, then all three doors would then have true statements which we know is not possible since one of them must be false.
2. If you assume Freedom is behind the middle door, then all three doors would then have false statements which we know is not

possible since one of them must be true.

3. Freedom is therefore behind door #3. The door #2 and door #3 both have true statements and door one has a false statement.

## 54. The Three Light Switches

**ANSWER**: Turn Switch 1 on and leave it on for a little while... about five minutes or more... and then turn it off. Turn Switch 2 on and go upstairs to inspect the lamps.

- The lamp with the bulb that is *off* but still *warm* is controlled by Switch 1.
- The lamp that is currently *on* is controlled by Switch 2.
- The lamp that is *off* and *cold* is controlled by Switch 3.

## 55. The Multiply Question

**ANSWER:** 1089 * 9 = 9801    A=1, B=0, C=8, D=9

## 56 Einstein's Riddle

**ANSWER**

The German sits in his Green House, smoking his Prince cigars, drinking coffee, and watching his FISH.

The rest go like this-
1st House: Yellow, Norwegian, Water, Cats, Dunhill
2nd House: Blue, Dane, Tea, Horse, Blends
3rd House: Red, Brit, Milk, Birds, Pall Malls
4th House: Green, German, Coffee, FISH, Prince
5th House: White, Swede, Beer, Dogs, Bluemasters

# Further reading and studying on your own...

This book attempts to focus on the front end RTL and Verilog interview questions. However, there are so many other topics which were not covered. You should continue to research and study on your own for these topics, or visit the website for information: **www.VerilogCode.com**

## Verification

- What are the differences between SOC level and IP level testing
- Differences between writing CPU compiled tests in C or assembly compared to unit level testing using constrained random tests
- Defining test plans and functional testing requirements
  - writing functional cross coverage buckets
- Measuring code coverage
  - line, branch, condition, and toggle coverage

## DFT (Design For Test)

- Understand scan flops and basics about how scan chains work

- Understand test modes for clock gating cells, latches (make them transparent), and memory (drive inputs to the outputs)

- Understand memory BIST or PBIST testing

## Physical Design Questions

As a front end RTL designer, you may not be familiar with STA or the physical design flow. However, you should be aware of what physical designers can do, and what methods and tools that they use. Including:

- Give an example of what is included in a synthesis constraint file
- If STA results are clean, but the netlist and sdf files are showing Xs in gate level simulations, what could be the possible causes?
  - incorrect constraint files
  - the physical placement (location) of cells

- clock domain crossings
- Understand some ways physical design can fix setup and hold violations, by:
  - Replication of logic
  - Resizing cells
  - Moving logic from right to left side
  - Breaking paths (pipeline techniques)
  - Under (PVT) relationships: Power, Voltage and Temperature
- Understand Dynamic Frequency Voltage Scaling

# Your Personal Interview Notes and Questions

Use these next pages for your own personal interview notes and experiences. As soon as you finish an interview, you should write down the questions immediately before you forget them. These pages can be used as your personal journal that you can review before each future interview.

From my experience, if you immediately write down the questions and then review all of them before your next interview then you will become more and more comfortable answering questions. It also builds up your confidence and library of information. Also, there are good chances that you may hear a related question on your next interview.

# Your Personal Interview Notes and Questions

# Your Personal Interview Notes and Questions

# Your Personal Interview Notes and Questions

# Your Personal Interview Notes and Questions

# Your Personal Interview Notes and Questions

# Your Personal Interview Notes and Questions

# Your Personal Interview Notes and Questions

# Your Personal Interview Notes and Questions

# Your Personal Interview Notes and Questions

# Your Personal Interview Notes and Questions

# Your Personal Interview Notes and Questions

# Your Personal Interview Notes and Questions

# Your Personal Interview Notes and Questions

# Credits and Sources

**Websites Sources:**

https://en.wikipedia.org/wiki/CMOS

http://www.design-reuse.com/articles/20775/hdl-design-low-power.html

http://www.asic-world.com

http://6004.mit.edu/Fall14/tutprobs/fsm.html

**White Papers:**

OSR Journal of VLSI and Signal Processing (IOSR-JVSP). ISSN: 2319 – 4200, ISBN No. : 2319 – 4197 Volume 1, Issue 3 (Nov. - Dec. 2012), PP 32-37 Asynchronous FIFO Design with Gray code Pointer for High Speed AMBA AHB Compliant Memory controller G.Ramesh, V.Shivaraj Kumar, K.Jeevan Reddy

Clock Dividers Made Easy by Mohit Arora. www.mikrocontroller.net

**People**

I would like to thank and acknowledge the following people who helped me during my preparation for my own job interviewing experience: Atif Hussain, Rajitha Padakanti, Bob Mizell, and William Wallace.

**Places**

I would like to thank the Red Horn Coffee House and Brewing Co. in Cedar Park, TX for providing a great atmosphere to work while providing some of the best coffee and craft beers in the Austin area.